Tolley's Practical Risk Assessment Handbook

Tolley's
Practical Risk Assessment Handbook

5th Edition

Mike Bateman
BSc, MIOSH, RSP
Health and Safety Consultant

AMSTERDAM • BOSTON • HEIDELBERG • LONDON • NEW YORK • OXFORD
PARIS • SAN DIEGO • SAN FRANCISCO • SINGAPORE • SYDNEY • TOKYO

Newnes is an imprint of Elsevier

Newnes is an imprint of Elsevier
Linacre House, Jordan Hill, Oxford OX2 8DP, UK
30 Corporate Drive, Suite 400, Burlington, MA 01803, USA

First edition 1999
Second edition 2000
Third edition 2002
Fourth edition 2003
Fifth edition 2006

British Library Cataloguing in Publication Data
A catalogue record for this book is available from the British Library

Library of Congress Cataloging in Publication Data
A cataloge record for this book is available from the Library of Congress

ISBN-13: 978-0-75-066989-4
ISBN-10: 0-75-066989-6

For information on all Newnes publications visit our website at http://books.elsevier.com

Typeset by Integra Software Services, Pvt. Ltd, Pondicherry, India
www.integra-india.com
Printed and bound in Great Britain

06 07 08 09 10 10 9 8 7 6 5 4 3 2 1

Contents

Preface ix

Foreword x

1 **Introduction** 1
The background to risk assessment 1
Regulations requiring risk assessment 6
Related health and safety management concepts 10
References 12

2 **What the Management Regulations require** 14
Introduction 14
Hazards and risks 16
Evaluation of precautions 17
Suitable and sufficient 18
Who should carry out the assessment? 20
Reviewing risk assessments 22
Related requirements of the Management Regulations 1999 24
References 30

3 **Special cases** 31
Children and young persons 32
Checklist: Work presenting increased risks for children and young persons 35
Prohibitions on children and young persons 40
New and expectant mothers 41
References 54

4 **Carrying out risk assessments** 55
Planning and preparation 55
Checklist of possible risks 63
Making the risk assessment 66
After the assessment 70
References 72

5 **Assessment records** 73
Introduction 73
Contents of assessment records 74
Sample assessment format 74
Illustrative assessments 74
Alternative assessment record formats 88

6 Model risk assessments 94
Introduction 94
Implementation and adaptation 94
Sample model assessments 95

7 Specialised risk assessment techniques 104
Introduction 104
Risk rating matrices 105
More sophisticated approaches 107
HAZOP 108

8 Implementation of precautions 112
Introduction 112
Principles of prevention 113
The management cycle 114
Planning 116
Organisation 117
Control 118
Monitoring 120
Review 130
References 132

9 COSHH Assessments 133
Introduction 134
How substances cause harm 134
The COSHH Regulations summarised 136
Planning and preparing for the assessment 138
Prevention or control of exposure 143
Making the assessment 149
Sample assessment records 152
After the assessment 156
Some pitfalls 169
References 170

10 Noise assessment 172
Introduction 173
How noise damages hearing 173
Noise measurement 174
The Control of Noise at Work Regulations 2005 summarised 176
Noise exposure reduction 179
Planning and preparation 184
Making the assessment 186
After the assessment 191
References 193

11 Assessment of manual handling 194
Introduction 194
What the Regulations require 195
Risk of injury from manual handling 198

	Avoiding or reducing risks	205
	Planning and preparation	213
	Making the assessment	215
	After the assessment	219
	References	223
12	**Assessment of DSE workstations**	224
	Introduction	224
	The Regulations summarised	225
	Planning and preparation	228
	Guidance on completing the DSE workstation self-assessment	235
	Special situations	237
	After the assessment	238
	References	239
13	**Assessment of Personal Protective Equipment (PPE) requirements**	240
	Introduction	240
	The Regulations summarised	241
	PPE assessment in practice	243
	Sample PPE assessment records	248
	References	251
14	**Fire and DSEAR assessments**	252
	Introduction	253
	Regulatory Reform (Fire Safety) Order 2005	253
	Planning and preparation for fire risk assesments	256
	Fire risk assessment factors	259
	Making the assessment	266
	After the assessment	271
	Dangerous Substances and Explosive Atmospheres Regulations 2002 (DSEAR)	273
	References	278
15	**Assessment of risks from asbestos in premises**	280
	Introduction	280
	The risks from asbestos	281
	Control of Asbestos at Work Regulations 2002	282
	Regulation 4 outlined	284
	Assessing the risks from asbestos	288
	The asbestos risk management plan	293
	References	300
16	**Assessment of work at height**	301
	Introduction	301
	The Regulations summarised	302
	Avoidance of risks from work at height	304
	Work equipment for work at height	304
	Use of ladders	307

	Planning and preparation	309
	Making the assessment	309
	Recording the assessment	310
	Dynamic risk assessments	313
	References	313
17	**Risk assessment related concepts**	315
	Introduction	315
	Safe systems of work	316
	Dynamic risk assesments	320
	Permits to work	321
	CDM health and safety plans	327
	Method statements	330
	References	332
18	**Assessing and managing risk – can you afford not to?**	333
	Introduction	333
	The Costs of Accidents at Work	334
	Some high-profile examples	335
	The benifit of assessing and managing risk well	336
	The potential costs of failure	338
	References	342
19	**Looking ahead**	343
	Risk assessment is here to stay	343
	Possible changes	344
	Widening the concept of work-related risks	345
	References	353
Index		354

Preface

This handbook is intended to help not only those involved in carrying out risk assessments but also those who simply need to know what risk assessment is all about – health and safety specialists, safety representatives, managers at all levels and directors, together with owners of small- and medium-sized businesses.

Whilst inevitably references must be made to legal requirements, the handbook adopts a practical approach to risk assessment and contains a variety of checklists, risk assessment forms and examples of what completed risk assessments might look like. It aims to remove much of the mystique about risk assessment and demonstrates that it is a process in which everyone with an awareness of the risks and precautions associated with their work activities can play a part.

This fifth edition takes account of several important changes in legislation since the previous edition.

1 **CHAPTER 16: ASSESSMENT OF WORK AT HEIGHT** is a completely new chapter which summarises the requirements of the *Work at Height Regulations 2005*. It explains the hierarchical approach which must be followed when assessing risks from work at height and the criteria which must be taken into account when selecting equipment for use in such work. Whether or not ladders should be used for work at height is also covered.

2 The important requirements of the Regulatory Reform (Fire Safety) Order have been included in **CHAPTER 14: FIRE AND DSEAR ASSESSMENTS**. This Order replaces previous legislation dealing with fire safety in the workplace.

3 **CHAPTER 10: NOISE ASSESSMENT** has been extensively modified to take account of the *Noise at Work Regulations 2005*, with the major changes since the previous regulations being highlighted.

The penalties being imposed by the courts for breaches of health and safety legislation have increased considerably. During 2005 two separate cases resulted in fines of £15 million and £10 million for single offences by employers. This trend is referred to in **CHAPTER 18: ASSESSING AND MANAGING RISK – CAN YOU AFFORD NOT TO?** An important question indeed!

Mike Bateman

Foreword

This updated and revised practical handbook is essential reading for all those engaged in and involved with workplace risk assessments, as it clearly and concisely outlines what is expected and required from a best practice viewpoint. Full use is made of practical risk assessment examples to enable the reader to see exactly what is required, and – more importantly – how to implement the findings that (should) result from risk assessments.

The revised handbook contains a variety of useful checklists and risk assessment formats, all having been the subject of successful implementation in the workplace. The updated fifth edition includes all the important, recent changes in the legislative requirements for both general and specific risk assessments such as for work at height and noise. It also includes a section dealing with the updated requirements for fire risk assessments under the Fire Safety Order.

As with previous editions, this excellent publication is a 'must' for directors, managers, health and safety professionals, safety representatives and others involved in risk assessments, as it gives practical advice and guidance on practical risk assessments.

It does just what it says on the tin!

Lawrence Bamber
B.Sc, DIS, CFIOSH, FIRM, MASSE
Past President of IOSH

1 Introduction

In this chapter:

The background to risk assessment **1.1**
Key requirements of HSWA 1974 **1.2**
What is 'reasonably practicable'? **1.3**
Practicable and absolute requirements **1.4**
The Management of Health and Safety at Work Regulations (SI 1999 No 3242) **1.5**

Regulations requiring risk assessment **1.6**
Control of Substances Hazardous to Health Regulations 2002 (SI 2002 No 2677) (COSHH 2002) **1.7**
Control of Noise at Work Regulations 2005 (SI 2005 No 1643) **1.8**
Manual Handling Operations Regulations 1992 (SI 1992 No 2793) **1.9**
Health and Safety (Display Screen Equipment) Regulations 1992 (SI 1992 No 2792) **1.10**
Personal Protective Equipment at Work Regulations 1992 (SI 1992 No 2966) **1.11**
Regulatory Reform (Fire Safety) Order 2005 **1.12**
Dangerous Substances and Explosive Atmospheres Regulations 2002 (SI 2002 No 2776) (DSEAR) **1.13**
Control of Asbestos at Work Regulations 2002 (SI 2002 No 2675) **1.14**
Work at Height Regulations 2005 (SI 2005 No 735) **1.15**
Control of Lead at Work Regulations 2002 (SI 2002 No 2676) **1.16**
Control of Asbestos at Work Regulations 2002 (SI 2002 No 2675) **1.17**
Supply of Machinery (Safety) Regulations 1992 (SI 1992 No 3073) **1.18**
Control of Major Accident Hazard Regulations 1999 (SI 1999 No 743) (COMAH 1999) **1.19**
Ionising Radiations Regulations 1999 (SI 1999 No 3232) **1.20**
Control of Vibration at Work Regulations 2005 (SI 2005 No 1093) **1.21**

Related health and safety management concepts **1.22**
Safe systems of work **1.23**
Dynamic risk assessment **1.24**
Permits to work **1.25**
CDM health and safety plans **1.26**
Method statements **1.27**

References **1.28**

The background to risk assessment

1.1 The term 'risk assessment' probably first came into common use as a result of the *Control of Substances Hazardous to Health Regulations 1988* (commonly known as the *'COSHH Regulations'* and revised several times since) which required employers to make a 'suitable and sufficient assessment of the risks created' by work liable to expose any employees to any substance hazardous to health. Similar requirements had actually previously been contained in both the *Control of Lead at Work Regulations 1980* and the *Control of Asbestos at Work Regulations 1987*.

In practice a type of risk assessment had already been necessary for some years particularly as a result of the use of the qualifying clause 'so far as is reasonably practicable' in a number of sections within the *Health and Safety at Work etc Act 1974* (*HSWA 1974*).

Key requirements of HSWA 1974

1.2 Previously most Acts and Regulations dealing with health and safety had been extremely prescriptive in their requirements and narrow in their scope. The report of the Robens Committee (published in 1972) recommended major changes including:

- the replacement of specific legal requirements by general obligations;
- legislation to cover everyone at work, including the self-employed (rather than just those in factories, offices, etc. as was previously the case);
- introduction of requirements for employers (and the self-employed) to take account not just of employees but also of others, including the public.

These recommendations were implemented through the passing of *HSWA* in 1974. The Act applies to everyone 'at work' – employers, self-employed and employees (with the exception of domestic servants in private households).

Section 2 sets out general duties of employers to their employees, with the most general contained in *Section 2(1)*:

> It shall be the duty of every employer to ensure, so far as is reasonably practicable, the health, safety and welfare at work of all his employees.

This 'catch-all' requirement is all-embracing in its scope, although it is qualified by the term 'reasonably practicable'. *Section 2(2)* goes on to detail more specific requirements relating to:

- provision and maintenance of plant and systems of work;
- use, handling, storage and transport of articles and substances;
- provision of information, instruction, training and supervision;
- places of work and means of access and egress;
- the working environment, facilities and welfare arrangements.

These are also qualified by the term 'reasonably practicable'.

Section 2(3) requires employers with five or more employees to prepare a written health and safety policy statement, together with the organisation and arrangements for carrying it out, and to bring this to the notice of employees.

Section 3 places general duties on both employers and the self-employed in respect of persons other than their employees. *Section 3(1)* states:

> It shall be the duty of every employer to conduct his undertaking in such a way as to ensure, so far as is reasonably practicable, that persons not in his employment who may be affected thereby are not exposed to risks to their health or safety.

Employers thus have duties to contractors (and their employees), visitors, customers, members of the emergency services, neighbours, passers-by and the public at large. This may (up to a point) extend to include trespassers. Once again these duties are subject to the 'reasonably practicable' qualification.

Self-employed persons are put under a similar duty by *Section 3(2)* and must also take care of themselves. (If they have employees then *Section 2* will also apply to them.)

Section 4 of the Act places duties on each person who has to any extent control of non-domestic premises used for work purposes in respect of those who are not their employees. Such persons may include landlords or managing agents. Each such person is required by *Section 4(2)*:

> to take such measures as it is reasonable for a person in his position to take to ensure, so far as is reasonably practicable, that the premises, all means of access thereto or egress therefrom available for use by persons using the premises, and any plant or substance in the premises or, as the case may be, provided for use there, is or are safe and without risks to health.

This requirement also is subject to the qualification of 'so far as is reasonably practicable'.

Section 6 places a number of duties on those who design, manufacture, import or supply articles for use at work, or articles of fairground equipment and those who manufacture, import or supply substances. Many of these obligations also contain the 'reasonably practicable' qualification. (It is not the intention of this handbook to develop further upon the duties contained in *Section 6*.)

What is 'reasonably practicable'?

1.3 The phrase 'reasonably practicable' is not just included in the key sections of *HSWA 1974* but is contained in many requirements of a wide variety of Regulations. Lord Justice Asquith provided a definition in his judgment on the case of *Edwards v. National Coal Board* (1949) in which he stated:

> 'Reasonably practicable' is a narrower term than 'physically possible' and seems to me to imply that a computation must be made by the owner in

> which the quantum of risk placed on one scale and the sacrifice involved in the measures necessary for averting risk (whether in money, time or trouble) is placed in the other, and that, if it be shown that there is a gross disproportion between them – the risk being insignificant in relation to the sacrifice – the defendants discharge the onus on them. Moreover, this computation falls to be made by the owner at a point in time anterior to the accident.

Section 40 of *HSWA 1974* places the burden of proof in respect of what was or was not 'reasonably practicable' (or 'practicable' – see below) on the person charged with failure to comply with a duty or requirement. To a certain extent this reverses the usual onus of proof although defendants need only establish that they satisfied the duty or requirement on the balance of probabilities (not beyond all reasonable doubt, as is normally the case in criminal courts).

Quite clearly it would be both impractical and undesirable to decide what precautions are appropriate for every work situation by making submissions to a court of law. Employers (and the self-employed) must make their own decisions – this is what the Robens report proposed – a large degree of self-regulation by employers rather than increasingly complex and specific legal requirements. The level of risk must be established and the various options as to precautions considered by the duty holder in order to determine what is 'reasonably practicable' – in effect the employer must carry out a form of risk assessment.

Two aspects of 'reasonably practicable' merit further emphasis. For a precaution not to be reasonably practicable, the risk must be insignificant in relation to the sacrifice involved in taking the precaution. The abilities of the duty holder to meet the sacrifice involved in averting the risk should not be a consideration. In other words the precaution must represent reasonable value in health and safety terms – whether the duty holder can afford the cost (or time or trouble) should not be an issue. Taking account of employers' differing abilities to bear the costs of precautions would obviously lead to extremely inconsistent application of the law.

Practicable and absolute requirements

1.4 Not all health and safety law is qualified by the phrase 'reasonably practicable'. Some requirements must be carried out 'so far as is practicable'. 'Practicable' is a tougher standard to meet than 'reasonably practicable'. However, its meaning is different from 'physically possible'. The precautions must be possible in the light of current knowledge and invention (*Adsett v. K and L Steelfounders & Engineers Ltd* (1953)). Once a precaution is practicable it must be taken even if it is inconvenient or expensive. However, it is not practical to take precautions against a danger which is not yet known to exist (*Edwards v. National Coal Board* (1949)), although it may be practicable once the danger is recognised. As stated in **1.3** above, *Section 40* of *HSWA 1974* places the burden of proving what may or may not be practicable on the duty holder.

Many health and safety duties are subject to neither 'practicable' nor 'reasonably practicable' qualifications. These absolute requirements usually state that something 'shall' or 'shall not' be done. However, even these duties often contain other words which are subject to a certain amount of interpretation – 'suitable', 'sufficient', 'adequate', 'efficient', 'appropriate', etc.

Once again a duty holder cannot hope to determine whether requirements have been met 'so far as is practicable' or whether the precise wording of an absolute requirement has been complied with unless a proper evaluation of the risks and the effectiveness of the precautions has been made, i.e. a form of risk assessment has been carried out.

The Management of Health and Safety at Work Regulations (SI 1999 No 3242)

1.5 The *Management of Health and Safety at Work Regulations* (the *Management Regulations*) were introduced in 1992 and revised in 1999. They were intended to implement the European Framework Directive (89/391) on the introduction of measures to encourage improvements in the safety and health of workers at work. *Regulation 3* of the *Management Regulations* 1992 required employers and the self-employed to make a suitable and sufficient assessment of the risks to both employees and persons not in their employment (in the latter case arising out of or in connection with the employer's or self-employed person's undertaking). The purpose of the assessment being the identification of the measures needed 'to comply with the requirements and prohibitions imposed... by or under the relevant statutory provisions'. i.e., identifying what is needed to comply with the law.

Given the extremely broad obligations contained in *Sections 2, 3, 4* and *6* of *HSWA 1974*, all risks arising from work activities should be considered as part of the risk assessment process (although some risks may be dismissed as being insignificant). Compliance with more specific requirements of Regulations must also be assessed – whether these are absolute obligations or subject to 'practicable' or 'reasonably practicable' qualifications.

The requirement for risk assessment introduced in the *1992 Management Regulations* simply formalised what employers should have been doing all along – identifying what precautions they needed to take to comply with the law. (A separate obligation in the *Management Regulations* required effective arrangements to be made to ensure these precautions were actually implemented.) However, in addition, *Regulation 3* also required all employers with five or more employees to record the significant findings of their assessments, i.e. they had to be able to demonstrate that they had actually gone through a systematic risk assessment process.

The 1999 version of the *Management Regulations* contained similar requirements and also consolidated various amendments which had been made to the original

Regulations. The most significant of these were that special attention must be given during the risk assessment process to risks to:

- new and expectant mothers;
- children and young persons.

More detailed guidance on these aspects and on the whole process of risk assessment is contained later in the handbook but the process can be summarised as establishing:

1. What risks arise from the work activities;
2. What precautions are in place;
3. Whether those precautions are enough to comply with the law;
4. If not, what additional precautions need to be introduced.

Any such additional precautions must then be implemented (*Regulation 5* of the *Management Regulations* 1999 contains requirements relating to the effective implementation of precautions).

Regulations requiring risk assessment

1.6 An increasing number of codes of Regulations contain requirements for risk assessments. Several of these Regulations are of significance to a wide range of work activities and are dealt with in some detail elsewhere in the handbook. These are summarised below.

Control of Substances Hazardous to Health Regulations 2002 (SI 2002 No 2677) (COSHH 2002)

1.7 *Regulation 6* requires employers to make a suitable and sufficient assessment of the risks created by work liable to expose any employees to any substance hazardous to health and of the steps that need to be taken to meet the requirements of the Regulations.

Control of Noise at Work Regulations 2005 (SI 2005 No 1643)

1.8 Under *Regulation 5*, employers who carry out work which is liable to expose any employees to noise at or above a lower exposure limit value (defined in the regulations) must make a suitable and sufficient assessment of the risk from that noise. The assessment must identify the measures which need to be taken to meet the requirements of the regulations.

Manual Handling Operations Regulations 1992 (SI 1992 No 2793)

1.9 Employers are required by *Regulation 4* to make a suitable and sufficient assessment of all manual handling operations at work which involve a risk of employees being injured and to take appropriate steps to reduce the risk to the lowest level reasonably practicable. (They must avoid such manual handling operations if it is reasonably practicable to do so.)

Health and Safety (Display Screen Equipment) Regulations 1992 (SI 1992 No 2792)

1.10 *Regulation 2* requires employers to perform a suitable and sufficient analysis of display screen equipment (DSE) workstations for the purpose of assessing risks to 'users' or 'operators' as defined in the Regulations. Risks identified in the assessment must be reduced to the lowest extent reasonably practicable.

Personal Protective Equipment at Work Regulations 1992 (SI 1992 No 2966)

1.11 Under *Regulation 6* employers must ensure that an assessment is made to determine risks which have not been avoided by other means and identify personal protective equipment (PPE) which will be effective against these risks. The Regulations also contain other requirements relating to the provision of PPE; its maintenance and replacement; information, instruction and training; and the steps which must be taken to ensure its proper use.

Regulatory Reform (Fire Safety) Order 2005

1.12 From October 2006, the *Fire Safety Order 2005* replaces the *Fire Precautions (Workplace) Regulations* and various other legal requirements dealing with fire. Article 9 of the Order requires the responsible person (defined elsewhere within the Order) to '. . . make a suitable and sufficient assessment of the risks to which relevant persons are exposed for the purpose of identifying the general fire precautions he needs to take . . .'

Dangerous Substances and Explosive Atmospheres Regulations 2002 (SI 2002 No 2776) (DSEAR)

1.13 These Regulations replaced the *Highly Flammable Liquids and Liquefied Petroleum Gases Regulations 1972* and several other pieces of legislation. They apply to substances which have the potential to create risks from fire, explosion, exothermic reactions, etc. Employers must carry out risk assessments of activities involving dangerous substances and take appropriate control measures. Some types of control measures are specified in the Regulations.

Control of Asbestos at Work Regulations 2002 (SI 2002 No 2675)

1.14 The Regulations contain a new *duty to manage asbestos in non-domestic premises*. Duty holders (defined in the Regulations) must carry out an assessment of what asbestos-containing materials (ACMs) are known or are presumed to be present and also prepare a written plan to manage the risks which may be created by ACM.

Some Regulations requiring risk assessments are of rather more specialist application and have not been included in the handbook. However, references to key Health and Safety Executive (HSE) publications of relevance are included at the end of this chapter. These Regulations include:

Work at Height Regulations 2005 (SI 2005 No 735)

1.15 Although these regulations do not introduce a separate requirement for risk assessment, *Regulation 6* sets out a hierarchy of measures which must be considered when carrying out a risk assessment in respect of work at height. The regulations also contain several other requirements in respect of work equipment for work at height which must be taken into account during the risk assessment process.

Control of Lead at Work Regulations 2002 (SI 2002 No 2676)

1.16 The 1980 version of these Regulations provided a basis for the original *COSHH Regulations*. *Regulation 5* of the 2002 Regulations requires an assessment to be made to determine the nature and the degree of exposure of employees to lead. (*Regulation* 3 also requires the exposure of others to be taken into account.) An assessment of the adequacy of control measures is required by the *Management Regulations* (see **REF. 1**).

Control of Asbestos at Work Regulations 2002 (SI 2002 No 2675)

1.17 The 2002 Regulations also continue the previous requirements for the 'assessment of work which exposes employees to asbestos'. These are contained in *Regulation 6* which specifies several aspects which must be taken into account during the assessment. The remainder of the Regulations follow a similar pattern to the *COSHH Regulations* and there are other detailed requirements relating to plans of work, notification of work with asbestos, designated areas, labelling, etc. (see **REFS 2, 3** and **4**).

Supply of Machinery (Safety) Regulations 1992 (SI 1992 No 3073)

1.18 The Regulations include a variety of procedures which must be followed in assessing conformity of machinery with essential health and safety requirements set out in the Machinery Directive. This assessment must be co-ordinated by the 'responsible person' – normally the manufacturer or his representative. However, where employers import machinery directly from outside the EC or where they assemble machinery or parts to form 'relevant machinery', they too are likely to have duties under the Regulations (see **REFS 5** and **6**).

Control of Major Accident Hazard Regulations 1999 (SI 1999 No 743) (COMAH 1999)

1.19 The Regulations only apply to sites containing specified quantities of dangerous substances. They require the preparation of both on-site and off-site emergency plans with the objectives of:

- containing and controlling incidents so as to minimise their effects, and to limit damage to persons, the environment and property;
- implementing the measures necessary to protect persons and the environment from the effects of major accidents;
- communicating the necessary information to the public and to the emergency services and authorities concerned in the area;
- providing for the restoration and clean-up of the environment following a major accident.

Such plans can only be prepared utilising risk assessment techniques and the HSE have provided considerable guidance on the methodology to be followed and the parameters to be taken into account (see **REFS 7, 8** and **9**).

Ionising Radiations Regulations 1999 (SI 1999 No 3232)

1.20 *Regulation* 7 requires employers to carry out a risk assessment before commencing any new activity involving work with ionising radiation. The assessment must be:

(2) sufficient to demonstrate that—

(a) all hazards with the potential to cause a radiation accident have been identified; and

(b) the nature and magnitude of the risks to employers and other persons arising from those hazards have been evaluated.

Where such radiation risks are identified, all reasonably practicable steps must be taken to:

- prevent any such accident;
- limit its consequences should such an accident occur;
- provide employees with necessary information, instruction, training and equipment necessary to restrict their exposure.

An HSE booklet contains an Approved Code of Practice and Guidance on the Regulations (see **REF. 10**). This includes an explanation of the inter-relationship between the above requirements and the risk assessments required by the *Management Regulations 1999.*

Control of Vibration at Work Regulations 2005 (SI 2005 No 1093)

1.21 Under *Regulation 5*, employers must carry out an assessment of risks from vibration to his employees, including identification of measures that need to be taken to meet the requirements of the regulations. Several publications of value in carrying out such an assessment are listed in the references section of this chapter (see **REFS 11–14**).

Related health and safety management concepts

1.22 Risk assessment techniques are an essential part of many health and safety management concepts and several of these are explored in some detail within the handbook.

Safe systems of work

1.23 Employers are required under *Section 2(2)(a)* of *HSWA 1974* to provide and maintain 'systems of work that are, so far as is reasonably practicable, safe and without risks to health'.

Regulation 4(2) of the *Confined Spaces Regulations 1997* contains a similar requirement for safe systems of work. *Regulation 8(1)* of the *Lifting Operations and Lifting Equipment Regulations 1998* (LOLER) requires lifting operations to be carried out in a safe manner, the accompanying Approved Code of Practice (ACOP) referring to the need for a safe system of work in certain circumstances. A safe system of work can only established through the process of risk assessment.

Dynamic risk assessment

1.24 Even applying the formalised risk assessment approaches described in the handbook will be impossible for employers to take account in advance of every possible variable and eventuality in work activities. A degree of reliance must be placed on employees to make their own judgements in respect of health and safety. For example:

- Will a maintenance task at a remote location require too much stretching and carrying of equipment to be carried out safely from a ladder?
- Is it safe to plough a sloping field given the prevailing ground and weather conditions?
- What precautions need to be taken before going to survey a semi-derelict building in a deprived part of a city?

The term 'dynamic risk assessment' is often used to describe the process employees are expected to follow in such situations. However, it is essential that employers ensure that employees have the necessary knowledge and experience to make such judgements. The employer's 'generic risk assessments' must have identified the types of risks which might be present in these variable situations, established a framework of precautions (procedures, equipment, etc.) which are likely to be necessary and provided guidance on which precautions are appropriate for which situations.

Permits to work

1.25 A permit to work system is a formalised method for identifying a safe system of work (usually for a high-risk activity) and ensuring that this system is followed.

The Permit Issuer is expected to carry out a dynamic risk assessment of the work activity. The Permit Issuer should be more competent in identifying the risks and the relevant precautions than those carrying out the work, and in many situations the types of precautions required will have been identified in advance. Further information on permits to work is provided in **CHAPTER 17**.

CDM health and safety plans

1.26 A key component of the *Construction (Design and Management) Regulations 1994* ('CDM 1994') is the requirement for a health and safety plan. The framework of the plan is prepared initially by the 'Planning Supervisor' (who is appointed by the client) and the plan is then developed in much more detail by the 'Principal Contractor' for the project, who is also responsible for its implementation. Essentially this process requires a risk assessment in relation to the CDM project:

- What risks are likely to be involved in the project?
- What precautions are likely to be required to control these risks?
- How will these precautions be implemented?
- Are the precautions proving effective?
- If not, what improvements need to be made?

More details about CDM health and safety plans are available in **CHAPTER 17**.

Method statements

1.27 The term 'method statement' is being used increasingly, particularly in relation to construction work. Method statements usually involve a description of how a particular task or operation is to be carried out. In the context of a CDM health and safety plan the method statement should identify all the components of a safe system of work – arrived at through a process of risk assessment. However, the method statement may go much further – identifying specification standards for the work being carried out, or equipment being installed and providing details on materials being used.

References

(All HSE publications)

1.28

1	L132	Control of lead at work (2002).
2	L27	Work with asbestos which does not normally require a licence. *Control of Asbestos at Work Regulations 2002* ACOP (2002).
3	L28	Work with asbestos insulation, asbestos coating and asbestos insulating board. *Control of Asbestos at Work Regulations* 2002 ACOP (2002).
4	L11	A guide to the *Asbestos (Licensing) Regulations* 1983. Guidance on Regulations (1999).
5	INDG 270	Supplying new machinery. Advice to suppliers (1998) – free leaflet.

6	INDG 271	Buying new machinery. A short guide to the law (1998) – free leaflet.
7	L111	A guide to the *Control of Major Accident Hazard Regulations* 1999 (1999).
8	HSG 190	Preparing safety reports. *Control of Major Accident Hazard Regulations* 1999 (1999).
9	HSG 191	Emergency planning for major accidents. *Control of Major Accident Hazard Regulations* (1999).
10	L121	Work With Ionising Radiation. *Ionising Radiations Regulations* 1999 ACOP & Guidance (2000).
11	L140	Hand-arm vibration. *Control of Vibration at Work Regulations* 2005. Guidance on Regulations (2005).
12	INDG 175	Control the risks from hand-arm vibration. Advice for employers (2005).
13	L141	Whole-body vibration. *Control of Vibration at Work Regulations* 2005. Guidance on Regulations (2005).
14	INDG 242	Control back-pain risks from whole-body vibration. Advice for employers (2005).

2 What the Management Regulations require

In this chapter:

Introduction	**2.1**
Hazards and risks	**2.2**
Evaluation of precautions	**2.3**
Suitable and sufficient	**2.4**
Small businesses with few or simple hazards	**2.5**
Intermediate cases	**2.5**
Large and hazardous sites	**2.5**
Who should carry out the assessment?	**2.6**
Reviewing risk assessments	**2.7**
Related requirements of the Management Regulations 1999	**2.8**
Principles of prevention to be applied (*Regulation 4*)	**2.9**
Health and safety arrangements (*Regulation 5*)	**2.10**
Health surveillance (*Regulation 6*)	**2.11**
Health and safety assistance (*Regulation 7*)	**2.12**
Procedures for serious and imminent danger and for danger areas (*Regulation 8*)	**2.13**
Contacts with external services (*Regulation 9*)	**2.14**
Information for employees (*Regulation 10*)	**2.15**
Co-operation and co-ordination (*Regulation 11*)	**2.16**
Persons working in host employers' or self-employed persons' undertakings (*Regulation 12*)	**2.17**
Capabilities and training (*Regulation 13*)	**2.18**
References	**2.19**

Introduction

2.1 A general requirement for risk assessment is contained in the *Management of Health and Safety at Work Regulations 1999* (the *Management Regulations 1999*), *Regulation 3*. Changes to the original Regulations passed in 1992 mean that the risk assessment must now include fire risks and precautions (see **CHAPTER 14: FIRE AND DSEAR ASSESSMENTS**) and also risks to both young persons (under 18s) and new and expectant mothers (see **CHAPTER 3: SPECIAL CASES**). Other Regulations require more specific types of risk assessment, e.g. of hazardous substances COSHH, noise, manual handling operations, DSE, personal protective equipment, dangerous substances and asbestos in premises and work at height (these are dealt with in **CHAPTERS 9–16** of the handbook).

Regulation 3(1) of the *Management Regulations 1999* states:

> every employer shall make a suitable and sufficient assessment of:
>
> (a) the risks to the health and safety of his employees to which they are exposed whilst they are at work; and
>
> (b) the risks to the health and safety of persons not in his employment arising out of or in connection with the conduct by him of his undertaking,
>
> for the purpose of identifying the measures he needs to take to comply with the requirements or prohibitions imposed upon him by or under the relevant statutory provisions and by Part II of the Fire Precautions (Workplace) Regulations 1997.

Regulation 3(2) imposes similar requirements on self-employed persons.

Regulation 3(3) requires a risk assessment to be reviewed if:

- there is reason to suspect that it is no longer valid; or
- there has been a significant change in the matters to which it relates.

Regulation 3(4) requires a risk assessment to be made or reviewed before an employer employs a young person and *Regulation 3(5)* identifies particular issues which must be taken into account in respect of young persons (especially their inexperience, lack of awareness of risks and immaturity). Further requirements in respect of young persons are contained in *Regulation 19*.

Regulation 16 contains specific requirements on the factors which must be taken into account in risk assessments in relation to new and expectant mothers. These relate to processes, working conditions and physical, biological or chemical agents. The 'special cases' of young persons and new or expectant mothers are dealt with in **CHAPTER 3: SPECIAL CASES**.

Regulation 3(6) requires employers with five or more employees to record:

- the significant findings of their risk assessments; and
- any group of employees identified as being especially at risk.

Many different methods of recording assessments are described in **CHAPTER 5: ASSESSMENT RECORDS.**

The HSE have published a booklet (*L21 Management of health and safety at work. Management of Health and Safety at Work Regulations 1999. Approved code of practice (2000)*) which contains the *Management Regulations* in full, the associated ACOP and guidance on the Regulations.

Hazards and risks

2.2 The ACOP to the Regulations provides definitions of both hazard and risk.

A *hazard* is something with the potential to cause harm.

A *risk* is the likelihood of potential harm from that hazard being realised.

The *extent of the risk* will depend on:

- the likelihood of that harm occurring;
- the potential severity of that harm (resultant injury or adverse health effect);
- the population which might be affected by the hazard, i.e. the number of people who might be exposed;

Example 1: Pedestrians crossing a roadway

Hazard: A pedestrian being injured through being hit by traffic.

Extent of the risk is determined by factors such as:

- the volume of traffic;
- the number of pedestrians having to cross the road;
- the layout of the roadway (designated crossing points, warning signs, lighting, lines of visibility);
- the speed of the traffic;
- the nature of the traffic (pedal cycles will cause less injury than heavy goods vehicles);
- the capabilities and awareness of those crossing the road (children probably being at greater risk than adults).

Example 2: Falling objects during maintenance work

Hazard: A person at ground level being struck by a falling object during maintenance work.

Extent of the risk is determined by factors such as:

- the frequency of maintenance at elevated levels;
- the presence of persons in areas below;
- the numbers of such persons;
- the security of tools used in the maintenance work;
- measures taken to secure other items, e.g. equipment and components;
- the degree of care exercised by maintenance workers;
- the size and weight of objects which may fall;
- the potential distance objects might fall;
- the presence of openings in the working platform;
- the availability of edge protection;
- whether persons at ground level are wearing head protection.

Evaluation of precautions

2.3 The ACOP clearly states that the risk assessment involves 'identifying the hazards present . . . and evaluating the extent of the risks involved, taking into account existing precautions and their effectiveness.'

The evaluation of the effectiveness of precautions is an integral part of the risk assessment process. This is overlooked by some organisations who concentrate on the identification (and often the quantification) of risks without checking whether the intended precautions are actually being taken in the workplace and whether these precautions are proving effective.

Taking the previous examples to illustrate this principle:

In *Example 1: Pedestrians crossing a roadway*, the risk assessment must include consideration of whether:

- pedestrians are using available crossing points;
- warning signs are of adequate size and suitably positioned;

- lighting levels are adequate;
- speed limits are being observed.

Similarly in *Example 2: Falling objects during maintenance work*, the risk assessment must take account of whether:

- barriers and/or warning signs at ground level are in place and being respected;
- tool belts are used by maintenance employees;
- edge protection and precautions are in place on working platforms;
- those at ground level are wearing head protection, if required.

Suitable and sufficient

2.4 Risk assessments under the *Management Regulations 1999* (and several other Regulations) must be 'suitable and sufficient', but the phrase is not defined in the Regulations themselves. However, the ACOP to the Regulations states that 'The level of risk arising from the work activity should determine the degree of sophistication of the risk assessment.' The ACOP also states that insignificant risks can usually be ignored, as can risks arising from routine activities associated with life in general ('unless the work activity compounds or significantly alters those risks').

In practice, a risk can only be concluded to be insignificant if some attention is paid to it during the risk assessment process and, if there is any scope for doubt, it is prudent to state in the risk assessment record which risks are considered insignificant. Some examples of this are contained in **CHAPTER 5: ASSESSMENT RECORDS.**

Similarly the risk assessment process should identify which routine activities of life might be compounded or significantly altered by work activities. Going out of doors in winter weather (with its attendant rain, ice, snow or wind) may be considered to be a routine activity. However, the risks may be considerably greater for work activities such as:

- work on electrical equipment situated outdoors;
- driving a fork lift truck in an icy yard;
- agricultural or construction work in a remote location;
- maintenance activities on an exposed part of a manufacturing plant.

2.5 The ACOP states that increasingly sophisticated risk assessments will be required in higher risk situations but unfortunately provides little in terms of illustrations and examples (see **CHAPTER 5: ASSESSMENT RECORDS** for examples based on practical experience of risk assessment).

The ACOP statements can be summarised as:

Small businesses with few or simple hazards

- A straightforward process based on informed judgement and reference to guidance.
- Obvious hazards and risks can be addressed directly.
- No complicated process or skills required.

Intermediate cases

- Some areas might require specialist advice, knowledge or techniques.

Large and hazardous sites

- A much more sophisticated approach will be required, especially for 'complex or novel processes'.
- Other legislation may require detailed safety cases or reports, e.g. bulk storage or use of hazardous substances, large scale mineral extraction or nuclear plant.

Emphasis is also placed in the ACOP upon the need for risk assessments to consider both workers and members of the public who might be affected by the undertaking. It refers to railway companies needing to take into account rail workers (their own employees and others), passengers and others, such as level crossing users.

Similarly a construction company would need to consider risks to (and from):

- their employees;
- sub-contractors;
- visitors to their sites;
- passers by;
- possible trespassers on sites.

A residential care home should take into account risks to (and from):

- their staff;
- residents;
- visiting medical specialists;
- visitors to residents;
- visiting contractors.

Further guidance on categories of people who may need to be covered in the risk assessment process is provided in **CHAPTER 4: CARRYING OUT RISK ASSESSMENTS** – see **4.5: CONSIDER WHO MIGHT BE AT RISK** and **4.8: CHECKLIST OF POSSIBLE RISKS.**

Who should carry out the assessment?

2.6 *Regulation 3* of the *Management Regulations 1999* requires the employer (or self-employed person) to make the assessment. However, *Regulation* 7 contains a requirement for employers to have competent health and safety assistance. *Paragraph (1)* of the Regulation states:

> Every employer shall, subject to paragraphs (6) and (7), appoint one or more competent persons to assist him in undertaking the measures he needs to take to comply with the requirements and prohibitions imposed upon him by or under the relevant statutory provisions and by Part II of the *Fire Precautions (Workplace) Regulations 1997*.

The 1999 version of the *Management Regulations* demonstrates a clear preference for the source of health and safety advice to be in the employer's employment, through *Paragraph (8)* which states:

> where there is a competent person in the employer's employment, that person shall be appointed for the purposes of paragraph (1) in preference to a competent person not in his employment.

Consequently, employers may appoint someone to carry out risk assessments on their behalf with the *Management Regulations 1999* expressing a preference for an employee over external persons such as health and safety consultants.

Paragraph (5) of the Regulation provides a definition of competence as:

> a person shall be regarded as competent for the purposes of paragraphs (1) and (8) where he has sufficient training and experience or knowledge and other qualities to enable him properly to assist in undertaking the measures referred to in paragraph (1).

Paragraphs (2) and (3) introduce requirements in respect of co-operation and the adequacy of the health and safety assistance.

> (2) Where an employer appoints persons in accordance with paragraph (1), he shall make arrangements for ensuring adequate co-operation between them.
>
> (3) The employer shall ensure that the number of persons appointed under paragraph (1), the time available for them to fulfil their functions and the means at their disposal are adequate having regard to the size of his undertaking, the risks to which his employees are exposed and the distribution of those risks throughout the undertaking.

Employers are also required by *Paragraph (4)* to ensure those providing health and safety assistance are given all necessary information about health and safety factors associated with their undertaking and those persons working in it.

It is also open to employers to act as their own source of health and safety assistance and carry out their own risk assessments. *Regulation 3, Paragraphs (6)* and (7) state:

(6) Paragraph (1) shall not apply to a self-employed employer who is not in partnership with any other person where he has sufficient training and experience or knowledge and other qualities to enable him properly to undertake the measures referred to in that paragraph himself.

(7) Paragraph (1) shall not apply to individuals who are employers and who are together carrying on business in partnership where at least one of the individuals concerned has sufficient training and experience or knowledge and other qualities–

(a) properly to undertake the measures he needs to take to comply with the requirements and prohibitions imposed upon him under the relevant statutory provisions; and

(b) properly to assist his fellow partners in undertaking the measures they need to take to comply with the requirements and prohibitions imposed upon them by or under the relevant statutory provisions.

The HSE leaflet *INDG 163 Five steps to risk assessment (1998)* states 'If you are a small firm and you are confident you understand what's involved, you can do the assessment yourself (you don't have to be a health and safety expert!).'

The key attributes are having 'sufficient training and experience or knowledge and other qualities' to be able to identify risks and evaluate the effectiveness of precautions to control those risks. Many people running small businesses should have such abilities in relation to their own work activities – this book aims to show them how best to use those abilities.

The HSE guidance to *Regulation* 7 refers to competence as not necessarily depending on the possession of particular skills or qualifications. It states that simple situations may only require:

- an understanding of relevant current best practice;
- an awareness of the limitations of one's own experience and knowledge; and
- the willingness and ability to supplement existing experience and knowledge, where necessary, by obtaining external help and advice.

Five steps to risk assessment suggests that larger firms ask a responsible employee, safety representative or safety officer to become involved in carrying out the risk assessment. However, the same leaflet recognises that help from external sources (such as health and safety consultants) may be necessary. A separate HSE leaflet *INDG 322 Need help on health and safety? (2000)* gives valuable guidance on this important subject. It makes reference to the different types of specialist support which are available from consultancies and also to the various qualifications available and the professional bodies active in the field of health and safety.

The Institution of Occupational Safety and Health (IOSH) is probably the pre-eminent professional body and will provide employers with guidance on consultancies available (IOSH, The Grange, Highfield Drive, Wigston, Leicestershire LE18 1NN, tel: 0116 257 3100).

Neither external consultants nor in-house health and safety specialists are likely to be able to conduct a risk assessment of a work activity without significant contact with those responsible for managing the activity, those who actually carry out the activity and also their health and safety representatives. Further advice on the importance of consultation is provided within **CHAPTER 4: CARRYING OUT RISK ASSESSMENTS** – see **4.11: DISCUSSIONS**.

Employers should be sceptical of consultancies and other organisations who offer to provide risk assessment documentation without the involvement of the employer and his staff. Such documents may be successful in identifying risks (particularly in more common work activities) and precautions required to control those risks. However, it is not usually possible to evaluate the effectiveness of the precautions without contact with those involved in the work activity and/or visiting the work location.

Reviewing risk assessments

2.7 *Regulation 3, Paragraph (3)* requires a risk assessment to be reviewed if:

(a) there is reason to suspect it is no longer valid;

(b) there has been a significant change in the matters to which it relates; and where as a result of any such review changes to an assessment are required, the employer or self-employed person concerned shall make them.

The ACOP to the *Management Regulations 1999* states that those carrying out risk assessments 'would not be expected to anticipate risks that were not foreseeable'. However, what is foreseeable can be changed by subsequent events. An accident, a non-injury incident or a case of ill-health may highlight the need for a risk assessment to be reviewed because:

- a previously unforeseen possibility has now occurred;
- the risk of something happening (or the extent of its consequences) is greater than previously thought;
- precautions prove to be less effective than anticipated.

Such information may come from outside the organisation itself – from others involved in the same work activity, thorough trade or specialist health and safety journals, from the suppliers of equipment or materials or from the HSE or other specialist bodies. Routine monitoring activity (inspections, audits, etc.) or consultation with employees may also identify the need for an assessment to be reviewed for the same sort of reasons as identified above. A review of the risk assessment may be required because of significant changes in the work activity – changes to the equipment or materials used, to the environment where the activity takes place, to the system of work used or to the numbers or types of people carrying out the activity.

A review of the risk assessment does not necessarily require a repeat of the whole risk assessment process but it is quite likely to identify the need for increased or changed precautions. However, in some cases the conclusion may be that even if the risks have increased, there are no precautions which are reasonably practicable available to control the risks.

Examples of changes requiring a review of a risk assessment (and the improved precautions which may be necessary) are:

- Larger components are to be processed on a machine tool (these necessitate larger openings in the guard, thus allowing easy access to dangerous parts)
  - extension tunnels need to be fitted to both sides of the guard,
  - improved mechanical handling apparatus is required.
- Increased storage needs mean that finished products are now stacked in the factory yard
  - stacks need to be protected from accidental impact from heavy goods vehicles,
  - fork lift truck drivers need to be provided with waterproof clothing.
- Night shift maintenance cover is reduced to one person
  - improved systems of communication with that person are required, e.g. portable phones, alarm devices, checks by security staff,
  - improved mechanical handling apparatus is required (for manual handling tasks previously carried out by two people),
  - a second person must be available from another source to assist in certain types of electrical work.

In practice, workplaces and the activities within them are constantly subject to gradual changes and the ACOP states 'it is prudent to plan to review risk assessments at regular intervals'. The frequency of such reviews should depend on the extent and the nature of the risks involved and the degree of change likely. In the author's view, all risk assessments should be reviewed at least every five years. Further guidance is provided in **CHAPTER 4: CARRYING OUT RISK ASSESSMENTS – see 4.20: ASSESSMENT REVIEW.**

As the ACOP points out, there are many activities where the nature of the work or the workplace itself changes constantly. Examples of such situations are construction work or peripatetic maintenance or repair work. Here it is possible to carry out 'generic' assessments of the types of risks involved and the types of precautions which should be taken. However, some reliance must be placed upon workers themselves to identify what precautions are appropriate for a given set of circumstances or to deal with unexpected situations. Clearly such workers must be well informed and well trained in order for them to be competent to make what are often called 'dynamic' risk assessments. This concept of dynamic risk assessment is dealt with in more detail in **CHAPTER 17: RISK ASSESSMENT RELATED CONCEPTS.**

Related requirements of the Management Regulations 1999

2.8 A number of other requirements of the *Management Regulations 1999* are of considerable importance to the process of risk assessment – they are part of the 'relevant statutory provisions' and the risk assessment must identify the measures necessary to comply with them.

Principles of prevention to be applied (Regulation 4)

2.9 *Regulation 4* and *Schedule 1* require preventive and protective measures to be implemented on the basis of specified principles. These principles will be referred to in more detail in **CHAPTER 8: IMPLEMENTATION OF PRECAUTIONS.**

Health and safety arrangements (Regulation 5)

2.10 Employers are required to have appropriate arrangements for the effective planning, organisation, control, monitoring and review of preventive and protective measures. Those employers with five or more employees must record these arrangements. This application of the 'management cycle' to health and safety matters will be covered in some detail in **CHAPTER 8: IMPLEMENTATION OF PRECAUTIONS.**

Health surveillance (Regulation 6)

2.11 Where risks to employees are identified through a risk assessment, *Regulation 6* requires that they 'are provided with such health surveillance as is appropriate'. In practice, surveillance is more likely to be necessary to comply with the requirements of more specific Regulations, e.g. COSHH, Noise, Asbestos, Ionising Radiations. However, some types of surveillance may be appropriate to deal with risks not covered elsewhere:

- colour blindness, e.g. in electricians or train drivers;
- other vision defects, e.g. in vehicle or train drivers;

- blackouts, epilepsy, e.g. in drivers, operators of machinery, those working at heights;
- capability of employees to carry out manual handling operations.

The HSE guidance is available on this important area (see *HSG 61 Health surveillance at work (1999)*).

Health and safety assistance (Regulation 7)

2.12 The requirements of this Regulation are described in **2.6: WHO SHOULD CARRY OUT THE ASSESSMENT?.**

Procedures for serious and imminent danger and for danger areas (Regulation 8)

2.13 *Regulation 8* requires every employer to 'establish and where necessary give effect to appropriate procedures to be followed in the event of serious and imminent danger to persons at work in his undertaking'. It also refers to the possible need to restrict access to areas 'on grounds of health and safety unless the employee concerned has received adequate health and safety instruction'.

The need for emergency procedures or restricted areas should of course be identified through the process of risk assessment. Situations 'of serious and imminent danger' might be due to fires, bomb threats, escape or release of hazardous substances, out of control processes, personal attack, escape of animals, etc.

The Regulation also requires the nomination of sufficient competent persons to implement evacuation (where this is necessary) and that emergency procedures should:

- 'so far as is practicable, require any persons at work who are exposed to serious and imminent danger to be informed of the nature of the hazard and of the steps taken or to be taken to protect them from it';
- enable such persons 'to stop work and immediately proceed to a place of safety in the event of their being exposed to serious, imminent and unavoidable danger';
- 'require the persons concerned to be prevented from resuming work in any situation where there is still a serious and imminent danger' (save in exceptional cases).

Areas may justify access to them being restricted because of the presence of hazardous substances, unprotected electrical conductors (particularly high voltage), potentially dangerous animals or people, etc.

Contacts with external services (Regulation 9)

2.14 This Regulation states that 'Every employer should ensure that any necessary contacts with emergency services are arranged, particularly as regards first aid, emergency medical care and rescue work.' Quite clearly this links very closely to the emergency procedures required by *Regulation 8*. The ACOP to the *Management Regulations 1999* provides further advice on the interface with external services.

Information for employees (Regulation 10)

2.15 Employers are required to provide employees with comprehensible and relevant information on:

- the risks to their health and safety identified by the risk assessment;
- the preventive and protective measures;
- emergency procedures.

This relates closely to the training requirements in *Regulation 13*. *Regulation 10* also includes specific requirements relating to the employment of children. These are dealt with in some detail in **CHAPTER 3: SPECIAL CASES.**

Co-operation and co-ordination (Regulation 11)

2.16 This Regulation requires employers and self-employed persons sharing a workplace (whether on a temporary or a permanent basis) to:

- co-operate with each other in respect of health and safety;
- co-ordinate their precautions;
- take all reasonable steps to inform each other about health and safety risks.

Sharing of workplaces may be temporary (such as in short-term construction work) or more permanent (as in a multi-occupancy building). Information may need to be provided to the other occupants about the use of hazardous substances or radioactive materials or in relation to high risk activities, particularly those which may affect the other occupants.

Co-ordination and co-operation are particularly likely to be necessary in relation to fire and other emergency procedures but may be relevant in other areas, e.g. leaving adequate space or visibility for vehicles used by other to circulate safely. Here, even those who do not have employees working on the premises but have responsibilities for common parts or services will also have an important part to play through their duties under *HSWA 1974, Section 4*.

Once again, the need to provide others with information should be identified through the risk assessment process. In respect of construction work, the requirements of the *Construction (Design and Management) Regulations 1994* (the *CDM Regulations 1994*) and particularly the need for a Health and Safety Plan must also be taken into account (see **CHAPTER 17: RISK ASSESSMENT RELATED CONCEPTS**).

Persons working in host employers' or self-employed persons' undertakings (Regulation 12)

2.17 Requirements are placed on host employers (or self-employed persons) to provide other employers (or self-employed persons) whose employees are working in the host's undertaking with comprehensible information on:

- risks to those visiting employees' health and safety arising out of or in connection with the conduct of the host's undertaking;
- precautions already taken by the host in relation to those visiting employees.

In addition, the host must 'ensure that any person working in his undertaking who is not his employee . . . is provided with appropriate instructions and comprehensible information regarding any risks to that person's health and safety which arise out of the conduct' of the host's undertaking. In both cases, this must include information about any evacuation procedures.

Essentially this Regulation identifies three issues which must be addressed during the risk assessment process.

Which visiting employees need to be considered?

Contractors are the obvious answer, but contractors provide a variety of services – cleaning, catering, security, servicing, maintenance, construction, etc. In some cases, workers supplied under a contract agreement may work alongside the host's own employees. Each of these groups will be exposed to different types of risks in relation to the host's undertaking. The Regulation can also be interpreted as applying to members of the emergency services who are exposed to risks (although no mention is made of this in the ACOP or HSE guidance on the Regulations).

What types of risk are relevant?

The Regulation itself places no limit on the risks in respect of which information must be provided. However, the ACOP indicates that the host should be concerned about the more significant or unusual risks associated with the host's activities or premises. The likely state of knowledge and awareness of the visiting employees may also need to be taken into account – an experienced service engineer may only need to be informed about risks peculiar to the host's activities or premises, whereas inexperienced workers supplied under contract are likely to need more comprehensive information.

How will the information be communicated?

The host has duties in respect of both the employer of visiting employees *and* the visiting employees themselves. Information (about risks, precautions and emergency procedures) might be provided to the employer together with the contract order. It might be in the form of a standard contractors' information sheet or handbook or through a document prepared specifically for the contractor in question. Documentation may need to be supported by face to face meetings and in some cases contractors' activities may be controlled through permit to work systems (see **CHAPTER 17: RISK ASSESSMENT RELATED CONCEPTS**).

Information can be supplied to visiting employees (about risks and emergency procedures) either directly by the host or via their own employer. However, it is the host's duty to ensure it is done. Depending on the nature and the extent of the risk, the host may choose to use an informal briefing or more formal induction methods (possibly even accompanied by a test to ensure the information has been assimilated). Again, use could be made of information sheets or contractors' handbooks, although the style and content may need to be different for visiting employees as opposed to their employers. It should be noted that these requirements overlap with some requirements of the *CDM Regulations 1994*, particularly the duties of the 'Principal Contractor' under those Regulations.

Detailed guidance on managing contractors and the *CDM Regulations 1994* is available from the HSE (*Managing contractors. A guide for employers (1997)*; *Managing health and safety in construction for health and safety. The Construction (Design and Management) Regulations 1994. Approved code of practice (2001)*; and *A guide to managing health and safety in construction (1995)*).

Capabilities and training (Regulation 13)

2.18 This Regulation is in three parts – *Paragraph (1)* states 'Every employer shall, in entrusting tasks to his employees, take into account their capabilities as regards health and safety.'

The capabilities of employees include not only the training they may have received but also their capacity to put that training into practice. Some tasks may require a high degree of skill in order to be carried out safely. In other situations, employees may be required to make a judgement about the risks involved and choose appropriate precautions, i.e. carry out a dynamic risk assessment. The employer must identify these types of tasks through the risk assessment process and be satisfied that staff performing them possess necessary skills or are properly equipped to make judgements on risks and precautions.

Paragraph (2) requires employers to ensure employees are provided with adequate health and safety training:

(a) on their being recruited into the employer's undertaking; and

(b) on their being exposed to new or increased risks because of

- a transfer or a change of responsibilities;
- introduction of new work equipment or changes in existing equipment;
- introduction of new technology; or
- a new or changed system of work.

Consequently the risk assessment should identify what should be included in general induction training (e.g. fire and other emergency procedures, general PPE requirements) and what training is required by employees carrying out specific tasks (e.g. operating equipment) or working in certain areas (e.g. high voltage switchrooms). Operating some types of equipment (e.g. a machine fully enclosed with fixed and interlocked guards) may only justify fairly simple health and safety training, whereas driving a fork lift truck, operating an overhead crane or using a chainsaw will require a much more comprehensive approach.

Employees who are transferred or given new responsibilities are also likely to need training in their new tasks or in relation to their new environment. Changes in equipment, technology or systems of work will necessitate a review of the risk assessment in relation to training needs, as well as other factors.

Paragraph (3) of the Regulation states:

> The training referred to in paragraph (2) shall –
>
> (a) be repeated periodically where appropriate;
>
> (b) be adapted to take account of new or changed risks to the health and safety of the employees concerned; and
>
> (c) take place during working hours.

The risk assessment should identify where refresher training may be necessary, e.g. where competence may decline if tasks are not performed regularly. Reviews of risk assessments may be prompted by changes in the workplace or by feedback from monitoring systems (audits, inspections, accident/incident investigations) and may indicate the need for changes to training programmes.

The ACOP to the Regulations states that 'if it is necessary to arrange training outside an employee's normal hours, this should be treated as an extension of time at work'. This would usually be expected to qualify for time off in lieu or overtime payments. Employers are required by the *Safety Representatives and Safety Committee Regulations (1977)* to consult safety representatives about arrangements for health and safety training. The ACOP emphasises that *HSWA 1974, Section 9* prohibits employers from requiring employees to pay for their own health and safety training.

Most of the other requirements of the *Management Regulations 1999* are of little relevance to the risk assessment process with the important exceptions of:

- risk assessment in respect of new or expectant mothers (*Regulation 16*); and
- protection of young persons (*Regulation 19*).

Both of these special cases are covered in detail in **CHAPTER 3: SPECIAL CASES.**

References

(All HSE publications)

2.19

1	L21	Management of health and safety at work. *Management of Health and Safety at Work Regulations 1999* Approved Code of Practice (2000)
2	INDG 163	Five steps to risk assessment (1998)
3	INDG 322	Need help on health and safety? (2000)
4	HSG 61	Health surveillance at work (1999)
5		Managing contractors. A guide for employers (1997)
6	HSG 224	Managing health and safety in construction. The *Construction (Design and Management) Regulations 1994* Approved Code of Practice and Guidance (2002)

3 Special cases

In this chapter:

Children and young persons 3.2
Introduction 3.2
Definitions 3.3
Requirements of the *Management Regulations 1999* 3.4
Assessing risks to children and young persons 3.5

Checklist: Work presenting increased risks for children and young persons 3.6
Excessively physically demanding work 3.6
Excessively psychologically demanding work 3.6
Harmful exposure to physical agents 3.6
Harmful exposure to biological or chemical agents 3.6
Work equipment 3.6
Dangerous processes or activities 3.6
Dangerous workplaces or workstations 3.6
Provision of information 3.8

Prohibitions on children and young persons 3.9
Carriage of explosives and dangerous goods 3.9
Agriculture 3.9
Ionising Radiation 3.9
Lead 3.9
Mines and quarries 3.9
Shipbuilding and shiprepairing 3.9
Docks 3.9

New and expectant mothers 3.10
Introduction 3.10
Requirements of the *Management Regulations 1999* 3.11
Risks to new or expectant mothers 3.12
Physical agents 3.13
Biological agents 3.14
Chemical agents 3.15
Working conditions 3.16
A practical approach to assessing risks 3.18

References 3.19

3.1 The *Management Regulations 1999* identify two particular types of workers to whom special attention must be given in carrying out risk assessments:

- children and young persons; and
- new and expectant mothers.

The requirements of the Regulations in respect of these categories are dealt with in this chapter.

Children and young persons

Introduction

3.2 Previous Acts and Regulations identified many types of equipment that children and young persons were not allowed to use, or processes or activities that they must not be involved in. Many of these 'prohibitions' were revoked by the *Health and Safety (Young Persons) Regulations 1997*, since incorporated into the *Management Regulations 1999*. (A few 'prohibitions' still remain – several of these are listed later in the chapter at **3.9: PROHIBITIONS ON CHILDREN AND YOUNG PERSONS**.) The emphasis has now changed to restrictions on the work that children and young persons are allowed to do, based upon the employer's risk assessment.

The *Health and Safety (Training for Employment) Regulations 1990* have the effect of giving students on work experience training programmes and trainees on training for employment programmes the status of 'employees'. The immediate provider of their training is treated as the 'employer'. (There are exceptions for courses at educational establishments, i.e. universities, colleges, schools, etc.) Therefore employers have duties in respect of all children and young persons at work in their undertaking – full-time employees, part-time and temporary employees and also students or trainees on work placement with them.

Definitions

3.3 The terms 'child' and 'young person' are defined in *Regulation 1(2)* of the *Management Regulations 1999*.

Child is defined as a person not over compulsory school age in accordance with:

- *the Education Act 1996, Section 8* (for England and Wales); and
- *the Education (Scotland) Act 1980, Section 31* (for Scotland).

(In practice this is just under or just over the age of sixteen.)

Young person is defined as 'any person who has not attained the age of eighteen'.

Some of the prohibitions remaining from older Health and Safety Regulations use different cut-off ages – where relevant to these are referred to later in the chapter at **3.9: PROHIBITIONS ON CHILDREN AND YOUNG PERSONS**. Further detail is also provided in HSE guidance (*HS(G) 165 Young people at work. A guide for employers (2000)* – see **REF. 1**).

The hours and types of work that children are allowed to do are subject to prohibitions and restrictions imposed by other (non-health and safety) legislation.

Requirements of the Management Regulations 1999

3.4 The general requirements for risk assessments placed on employers by *Regulation 3* of the *Management Regulations 1999* were described in the previous chapter (see **2.1: INTRODUCTION**). *Regulation 3, Paragraphs (4)* and *(5)* introduce specific requirements in respect of young persons, i.e. all under 18s.

> (4) An employer shall not employ a young person unless he has, in relation to risks to the health and safety of young persons, made or reviewed an assessment in accordance with paragraphs (1) and (5).
>
> (5) In making or reviewing the assessment, an employer who employs or is to employ a young person shall take particular account of–
>
> (a) the inexperience, lack of awareness of risks and immaturity of young persons;
>
> (b) the fitting-out and layout of the workplace and the workstation;
>
> (c) the nature, degree and duration of exposure to physical, biological and chemical agents;
>
> (d) the form, range and use of work equipment and the way in which it is handled;
>
> (e) the organisation of processes and activities;
>
> (f) the extent of the health and safety training provided or to be provided to young persons; and
>
> (g) risks from agents, processes and work listed in the Annex to Council Directive 94/33/EC on the protection of young persons at work.

The risk assessment in respect of young persons must also take particular note of the requirements of *Regulation 19* which states:

> (1) Every employer shall ensure that young persons employed by him are protected at work from any risks to their health or safety which are a consequence of their lack of experience, or absence of awareness of existing or potential risks or the fact that young persons have not yet fully matured.
>
> (2) Subject to paragraph (3), no employer shall employ a young person for work–
>
> (a) which is beyond his physical or psychological capacity;
>
> (b) involving harmful exposure to agents which are toxic or carcinogenic, cause heritable genetic damage or harm to the unborn child or which in any way chronically affect human health;
>
> (c) involving harmful exposure to radiation;
>
> (d) involving the risk of accidents which it may reasonably be assumed cannot be recognised or avoided by young persons owing to their insufficient attention to safety or lack of experience or training; or

(e) in which there is a risk to health from–

(i) extreme cold or heat,

(ii) noise, or

(iii) vibration;

and in determining whether work will involve harm or risk for the purposes of this paragraph, regard shall be had to the results of the assessment.

(3) Nothing in paragraph (2) shall prevent the employment of a young person who is no longer a child for work–

(a) where it is necessary for his training;

(b) where the young person will be supervised by a competent person; and

(c) where any risk will be reduced to the lowest level that is reasonably practicable.

(4) The provisions contained in the Regulations are without prejudice to–

(a) the provisions contained elsewhere in these Regulations; and

(b) any prohibition or restriction, arising otherwise than by this Regulation, on the employment of any person.

Assessing risks to children and young persons

3.5 A good starting point in assessing risks is to consider the three characteristics associated with young people which are mentioned in both *Regulations 3(5)* and *19(1)*:

- lack of experience;
- lack of awareness of existing or potential risks; and
- immaturity (in both the physical and psychological sense).

All young people share these characteristics but to differing extents – for example, one would have different expectations of a school leaver who had already been playing a prominent role in a family business such as a shop, restaurant or farm, as opposed to a work experience student or trainee with no previous exposure to the world of work. The degrees of physical and psychological maturity of young people also vary hugely.

These three characteristics must then be considered in respect of the risks involved in the employer's work activities and particularly those identified in *Regulations 3(5)* and *19(2)*. As an aid to the risk assessment process, the contents of those regulations (including those risks listed in the Annex to *Council Directive 94/33/EC*) have been consolidated into a single checklist, given in the next page.

Checklist: Work presenting increased risks for children and young persons

3.6 This checklist is based upon the *Management of Health and Safety at Work Regulations 1999, Regulations 3(5)* and *19(2)* and the Annex to *Council Directive 94/33/EC*. It is intended to assist employers conducting risk assessments in respect of work by children and young persons.

These types of work are not necessarily prohibited, although the requirements of *Regulation 19(2)* must be taken into account. However, such work is likely to require restrictions for most young persons (particularly children) and additional precautions are likely to be required to provide them with adequate protection from risk.

Excessively physically demanding work

- Manual handling operations where the force required or the repetitive nature of the activity could injure someone whose body is still developing (including production line work).
- Certain types of piecework.

Excessively psychologically demanding work

- Work with difficult clients or situations where there is a possibility of violence or aggression.
- Difficult emotional situations, e.g. dealing with death, serious illness or injury.
- Decision-making under stress.

Harmful exposure to physical agents

- Ionising radiation.
- Non-ionising radiation, e.g. lasers, UV from electric arc welding, IR from furnaces or burning/welding.
- Risks to health from extreme cold or heat.
- Excessive noise.
- Hand-arm vibration, e.g. from portable tools.
- Whole-body vibration, e.g. from off-road vehicles.
- Work in pressurised atmospheres and diving work.

Harmful exposure to biological or chemical agents

- Toxic or carcinogenic substances (including lead and asbestos).
- Substances causing heritable genetic damage or harming the unborn child.
- Substances chronically affecting human health.
- Other hazardous substances (harmful, corrosive, irritant).

Work equipment

Where there is an increased risk of injury due to the complexity of precautions required or the level of skill required for safe operation, for example:

- woodworking machines;
- food slicers and other food processing machinery;
- certain types of portable tools such as chainsaws;
- setting of power presses;
- vehicles such as fork lift trucks, mobile cranes, construction vehicles;
- firearms.

Dangerous processes or activities

- Work with explosives, including fireworks.
- Work with fierce or poisonous animals, e.g. on farms, in zoos or veterinary work.
- Certain types of electrical work, e.g. exposure to high voltage or live electrical equipment.
- Handling of highly flammable materials, e.g. petrol, other flammable liquids, flammable gases.
- Work with pressurised gases.
- Work in large slaughterhouses.
- Holding large quantities of cash or valuables.

Dangerous workplaces or workstations

- Work at heights, e.g. on high ladders or other unprotected forms of access.
- Work in confined spaces, particularly where the risks specified in the *Confined Spaces Regulations 1997* are present.
- Work where there is a risk of structural collapse, e.g. in construction or demolition activities or inside old buildings.

3.7 The purpose of all risk assessments is to identify what measures the employer needs to take in order to comply with the law. Measures which the employer must consider in order to provide adequate protection for young persons are:

- not exposing the young person to the risk at all;*
- providing additional training;
- providing close supervision by a competent person;
- carrying out additional health surveillance (as required by *Regulation 6* of the *Management Regulations 1999* or other Regulations, e.g. *COSHH 2002*);
- taking other additional precautions.

* *Regulation 19(2)* means that serious consideration must always be given to whether the risks remaining after control measures have been applied require this option to be taken. More flexibility is allowed by *Regulation 19(3)* in respect of young persons who are no longer children.

In deciding what precautions are required, the employer must consider both young people generally and the characteristics of individual young persons. For example, higher levels of training and/or supervision may be necessary in respect of young people with 'special needs'. Young people should gradually acquire more experience, awareness of risks and maturity, particularly as they pass through formal training programmes within the NVQ system. As this occurs, restrictions on their activities may be progressively removed.

Even when young workers have passed the age of 18, it should be noted that *Regulation 13* of the *Management Regulations 1999* still requires employers to take account of their capabilities in entrusting tasks to them and also to provide them with adequate health and safety training.

Provision of information

3.8 *Regulation 10* of the *Management Regulations 1999* contains general requirements for employers to provide all employees with information about risks and associated precautions. It must always be borne in mind that young persons will generally be less aware of risks than more experienced workers. However, important additional requirements in respect of the employment of children are contained in *Paragraphs (2)* and *(3)* of the Regulation which state:

(2) Every employer shall, before employing a child, provide a parent of the child with comprehensible and relevant information on–

(a) the risks to their health and safety identified by risk assessment;

(b) the preventive and protective measures; and

(c) the risks notified to him in accordance with *Regulation 11(1)(c)*.

(3) The reference in *paragraph (2)* to a parent of the child includes–

(a) in England and Wales, a person who has parental responsibility, within the meaning of *section 3* of the *Children Act 1989*, for him; and

(b) in Scotland, a person who has parental rights, within the meaning of *section 8* of the *Law Reform (Parent and Child) (Scotland) Act 1986*, for him.

This requirement to provide information to parents includes situations where children are on work experience programmes (where they have the status of employees by virtue of the *Health and Safety (Training for Employment) Regulations 1990)* and also includes part time or temporary work.

The wide-ranging impact of these provisions is lessened to a certain extent by the introduction of *Paragraph (2)* into *Regulation 2* 'Disapplication of these Regulations'. This states:

(2) Regulations 3(4), (5), 10(2) and 19 shall not apply to occasional work or short-term work involving–

(a) domestic service in a private household; or

(b) work regulated as not being harmful, damaging or dangerous to young people in a family undertaking.

However, the key term 'family undertaking' is not defined in the Regulations. The HSE guidance (*HS(G) 165 Young people at work. A guide for employers (2000), page 3*) indicates that this should be interpreted as meaning a firm owned by, and employing members of the same family, i.e. husbands, wives, fathers, mothers, grandfathers, grandmothers, stepfathers, stepmothers, sons, daughters, grandsons, granddaughters, stepsons, stepdaughters, brothers, sisters, half-brothers and half-sisters.

As far as is known, this narrow interpretation of 'family undertaking' has not been tested in the courts. Common usage of the term would suggest small- and medium-sized businesses controlled and managed by members of the same family but not necessarily only employing family members, as implied by the words used by the HSE.

The HSE guidance accompanying *Regulation 10* allows for the information on the key findings of the risk assessment to be passed on verbally to parents or guardians or even via the child itself (although it acknowledges the potential fallibility of this latter method). However, many would recommend the provision of a written summary – most organisations managing work experience programmes already have standard forms for doing this.

As the HSE acknowledge, the information required by *Regulation 10* to be given to all young persons may be provided either orally or in writing or possibly both. Key items (eg critical restrictions or prohibitions) should be recorded.

Means of providing information might include:

- induction training programmes;
- employee handbooks or rulebooks;

- job descriptions;
- formal operating procedures;
- trainee agreement forms (increasingly common for young persons on formal training programmes, e.g. modern apprenticeships);
- information forms for parents of work experience students.

The information must be comprehensible – special arrangements may be necessary for young people whose command of English is poor or for those with special needs. The type of information required to be provided will obviously relate to the work activities and the risks involved. The content must be relevant – to both the workplace and the young person. It might include the following types of information:

- general risks present in the workplace – e.g. fork lift trucks are widely used in the warehouse;
- general precautions taken in respect of those risks – e.g. all fork lift drivers are trained to the standard required by the ACOP;
- specific precautions in respect of the young person – e.g. the induction tour includes identification of areas where fork lift trucks operate and indication of warning signs;
- restrictions or prohibitions on the young person – e.g. X will not be allowed to drive fork lift trucks or any other vehicles (he will be considered for fork lift truck training after attaining the age of 17);
- supervision arrangements – e.g. X will be supervised by the warehouse foreman (or other persons designated by him);
- PPE (personal protective equipment) requirements – e.g. safety footwear must be worn by all employees working in the warehouse (this is supplied by the company).

Where restrictions or prohibitions are removed (e.g. after successful completion of training programmes), an appropriate record should be made, either on the original restriction/prohibition or upon the individual's training record.

Prohibitions on children and young persons

3.9 Outright prohibitions on work by children and young persons continue to be revoked, and to be replaced by a risk assessment approach. This is often supported by HSE guidance referring to circumstances in which certain activities may be acceptable for a young person to carry out, e.g. adequate training and supervision, necessary maturity and competence, etc. Prohibitions known to be still in place in 2000 (the year of the publication of the latest HSE guidance booklet) (**REF. 1**) were as follows.

Carriage of explosives and dangerous goods

Under-18s may not be employed as drivers or attendants of explosives vehicles, nor be responsible for the security of the explosives and may only enter the vehicle under the direct supervision of someone over the age of 18. (There are exceptions where the risks are slight.) [*Carriage of Explosives by Road Regulations 1996*]

Under-18s may not supervise road tankers or vehicles carrying dangerous goods nor supervise the unloading of petrol from a road tanker at a petrol-filling station. [*Carriage of Dangerous Goods by Road Regulations 1996* as amended]

Agriculture

Under-13s may not ride on vehicles and machines including tractors, trailers, etc. [*Prevention of Accidents to Children in Agriculture Regulations 1998*]

The HSE guidance also states that children (under the minimum school-leaving age) should not operate certain machines and tractors carrying out certain operations.

Ionising Radiation

Under-18s may not be designated as 'classified persons'. Dose exposure limits are lower for under-18s. [*Ionising Radiations Regulations 1999*]

Lead

Under-18s may not be employed in certain lead processes, e.g. lead smelting and refining, lead-acid battery manufacturing, nor should they be allowed to clean places where such processes are carried out.

Mines and quarries

Various restrictions exist for under-18s and under-16s relating to the use of winding and rope haulage equipment, conveyors at work faces, locomotives, shunting and shot firing. Some restrictions also apply to under-21s and under- 22s.

Shipbuilding and shiprepairing

In relation to under-18s, until they have been employed in a shipyard for six months, may not be employed on staging or in any part of a ship where they are liable to fall more than two metres or into water where there is a risk of drowning. [*Shipbuilding and Shiprepairing Regulations 1960*]

Docks

Under-18s may not operate powered lifting appliances in dock operations unless undergoing a suitable course of training under proper supervision of a competent person (serving members of HM Forces are exempt). [*Docks Regulations 1988*]

New and expectant mothers

Introduction

3.10 Amendments were made in 1994 to the previous *Management Regulations* implemented the *EC Directive on Pregnant Workers (92/85/EEC)* requiring employers in their risk assessments to consider risks to new or expectant mothers. These amendments were subsequently incorporated into the *Management Regulations 1999*. *Regulation 1* of the *Management Regulations 1999* contains two relevant definitions:

- *New or expectant mother* means an employee who is pregnant, who has given birth within the previous six months, or who is breastfeeding.
- *Given birth* means 'delivered a living child or, after twenty-four weeks of pregnancy, a stillborn child'.

Requirements of the Management Regulations 1999

3.11 The requirements for 'Risk Assessment in respect of new and expectant mothers' are contained in *Regulation 16* which states in *Paragraph (1)*:

Where–

(a) the persons working in an undertaking include women of child-bearing age; and

(b) the work is of a kind which could involve risk, by reason of her condition, to the health and safety of a new or expectant mother, or to that of her baby, from any processes or working conditions, or physical, biological or chemical agents, including those specified in Annexes I and II of Council Directive 92/85/EEC on the introduction of measures to encourage improvements in the safety and health at work of pregnant workers and workers who have recently given birth or are breastfeeding;

the assessment required by *Regulation 3(1)* shall also include an assessment of such risk.

Regulation 16, Paragraph (4) states that in relation to risks from infectious or contagious diseases an assessment must only be made if the level of risk is in addition to the level of exposure outside the workplace.

The types of risk which are more likely to affect new or expectant mothers are described later in the chapter at **3.12: RISKS TO NEW OR EXPECTANT MOTHERS**. *Regulation 16, Paragraphs (2)* and *(3)* set out the actions employers are required to take if these risks cannot be avoided. *Paragraph (2)* states:

> Where, in the case of an individual employee, the taking of any other action the employer is required to take under the relevant statutory provisions would not avoid the risk referred to in paragraph (1) the employer shall, if it is reasonable to do so, and would avoid such risks, alter her working conditions or hours of work.

Consequently where the risk assessment required under *Regulation 16(1)* shows that control measures would not sufficiently avoid the risks to new or expectant mothers or their babies, the employer must make reasonable alterations to their working conditions or hours of work.

In some cases, restrictions may still allow the employee to substantially continue with her normal work but in others it may be more appropriate to offer her suitable alternative work.

Any alternative work must be:

- suitable and appropriate for the employee to do in the circumstances;
- on terms and conditions which are no less favourable.

Regulation 16, Paragraph (3) states:

> If it is not reasonable to alter the working conditions or hours of work, or if it would not avoid such risk, the employer shall, subject to section 67 of the 1996 Act, suspend the employee from work for so long as is necessary to avoid such risk.

The '1996 Act' referred to here is the *Employment Rights Act 1996* which provides that any such suspension from work on the above grounds is on full pay. However, payment might not be made if the employee has unreasonably refused an offer of suitable alternative work. Employment continues during such a suspension, counting as continuous employment in respect of seniority, pension rights, etc. Contractual benefits other than pay do not necessarily continue during the suspension. These are a matter for negotiation and agreement between the employer and the employee, although employers should not act unlawfully under the *Equal Pay Act 1970* and the *Sex Discrimination Act 1975*. Enforcement of employment rights is through employment tribunals.

Regulation 17 of the *Management Regulations 1999* deals specifically with night work by new or expectant mothers and states:

> Where–
>
> (a) a new or expectant mother works at night; and
>
> (b) a certificate from a registered medical practitioner or a registered midwife shows that it is necessary for her health or safety that she should not be at work for any period of such work identified in the certificate,
>
> the employer shall, subject to section 46 of the 1978 Act, suspend her from work for so long as is necessary for her health or safety.

Such suspension (on the same basis as described above) is only necessary if there are risks arising from work. The HSE do not consider there are any risks to pregnant or breastfeeding workers or their children working at night per se. They suggest that any claim from an employee that she cannot work nights should be referred to an occupational health specialist. The HSE's own Employment Medical Advisory Service is likely to have a role to play in such cases.

The requirements placed on employers in respect of altered working conditions or hours of work and suspensions from work only take effect when the employee has formally notified the employer of her condition. *Regulation 18* states:

> (1) Nothing in paragraph (2) or (3) of *Regulation 16* shall require the employer to take any action in relation to an employee until she has notified the employer in writing that she is pregnant, has given birth within the previous six months, or is breastfeeding.

(2) Nothing in paragraph (2) or (3) of *Regulation 16* or in *Regulation 17* shall require the employer to maintain action taken in relation to an employee–

(a) in a case–

(i) to which Regulation 16(2) or (3) relates; and

(ii) where the employee has notified her employer that she is pregnant, where she has failed, within a reasonable time of being requested to do so in writing by her employer, to produce for the employer's inspection a certificate from a registered medical practitioner or a registered midwife showing that she is pregnant;

(b) once the employer knows that she is no longer a new or expectant mother; or

(c) if the employer cannot establish whether she remains a new or expectant mother.

Risks to new or expectant mothers

3.12 The HSE booklet *HS(G) 122 New and expectant mothers at work. A guide for employers* (**REF. 2**) provides considerable guidance on those risks which may be of particular relevance to new or expectant mothers, including those listed in the *EC Directive on Pregnant Workers (92/85/EEC)*. This guidance is both summarised and augmented below.

Physical agents

Manual Handling

3.13 Pregnant women are particularly susceptible to risk from manual handling activities as also are those who have recently given birth, especially after a caesarean section. Manual handling assessments (as required by the *Manual Handling Operations Regulations 1992*) are dealt with in some detail in **CHAPTER 11: ASSESSMENT OF MANUAL HANDLING**.

Noise

The HSE do not consider that there are any specific risks from noise for new or expectant mothers. Compliance with the requirements of the *Control of Noise at Work Regulations 2005* should provide them with sufficient protection.

Ionising radiation

The foetus may be harmed by exposure to ionising radiation, including that from radioactive materials inhaled or ingested by the mother. The *Ionising Radiations Regulations 1999* set an external radiation dose limit for the abdomen of any woman of reproductive capacity and also contain a specific requirement to provide information to female employees who may become pregnant or start breastfeeding. Systems of work should be such as to keep exposure of pregnant women to radiation from all sources as low as reasonably practicable. Contamination of a nursing mother's skin with radioactive substances can create risks for the child and special precautions may be necessary to avoid such a possibility.

Several HSE publications provide detailed guidance on work involving ionising radiation (see *L 121 Work with ionising radiation. Ionising Radiations Regulations 1999. Approved code of practice and guidance (2000)*; and *INDG 334 Working safely with radiation. Guidelines for expectant or breastfeeding mothers (2001)*, in particular).

Other electromagnetic radiation

The HSE do not consider that new or expectant mothers are at any greater risk from other types of radiation, with the possible exception of over exposure to radio-frequency radiation which could raise the body temperature to harmful levels. Compliance with the exposure standards for electric and magnetic fields published by the National Radiological Protection Board should provide adequate protection.

Work in compressed air

Although pregnant women may not be at greater risk of developing the 'bends', potentially the foetus could be seriously harmed by gas bubbles in the circulation should this condition arise. There is also evidence that women who have recently given birth have an increased risk of the bends.

The *Work in Compressed Air Regulations 1996, Regulation 16(2)* states:

> the compressed air contractor shall ensure that no person works in compressed air where the compressed air contractor has reason to believe that person to be subject to any medical of physical condition which is likely to render that person unfit or unsuitable for such work.

This would appear to prohibit such work by pregnant women or those who have recently given birth. In their booklet *HS(G)122*, the HSE state that there is no physiological reason why a breastfeeding mother should not work in compressed air although they point out that practical difficulties would exist. The HSE booklet *L96 A guide to the Work in Compressed Air Regulations 1996 (1996)* provides detailed guidance on the requirements of the *Work in Compressed Air Regulations 1996.*

Diving work

The HSE draw attention to the possible effects of pressure on the foetus during underwater diving by pregnant women and state that they should not dive at all.

Under the *Diving at Work Regulations 1997, Regulation 15,* divers must have a certificate of medical fitness to dive and the HSE guidance to doctors issuing such certificates advises that pregnant workers should not dive.

Regulation 13(1) states that 'No person shall dive in a diving project –. . . if he knows of anything (including any illness or medical condition) which makes him unfit to dive'.

A series of HSE booklets provide general guidance on different types of diving projects (see L 103 Commercial diving projects offshore (1998); L 104 Commercial diving projects inland/inshore (1998); L 105 Recreational diving projects (1998); L 106 Media diving projects (1998); *and* L 107 Scientific and archaeological diving projects (1998)).

Shock, vibration, etc.

Major physical shocks or regular exposure to lesser shocks or low frequency vibration may increase the risk of a miscarriage. Activities involving such risks (e.g. the use of vehicles off-road) should be avoided by pregnant women. Several sources of guidance on risks from vibration are listed in the References section of **CHAPTER 1**.

Movement and posture

Fatigue from standing and other physical work has been associated with miscarriage, premature birth and low birth weight. Ergonomic considerations will increase as the pregnancy advances and these could affect DSE workstations (discussed in **3.16: WORKING CONDITIONS**, below), work in restricted spaces (e.g. for some maintenance or cleaning activities), or work on ladders or platforms. Underground mining work is likely to involve movement and posture problems and will also be subject to some of the other 'physical agents' described in this section. Driving for extended periods or travel by air may also present postural problems.

Pregnant women should be allowed to pace their work appropriately, taking longer and more frequent breaks. They may need to be restricted from carrying out certain tasks. Seating may need to be provided for work that is normally done standing, and adjustments to DSE and other workstations may be necessary.

Physical and mental pressure

Excessive physical or mental pressure could cause stress and lead to anxiety and raised blood pressure. Workplace stress is a complex issue which has recently been receiving increased attention. Fatigue issues are referred to above but other possible causes of stress may also need to be considered. These might be associated with the workload of individual pregnant employees (e.g. for those in management or administrative roles), the pressure of decision-making (e.g. in the health care or financial sectors) or the trauma of certain possible work situations (e.g. in the emergency services).

Extreme cold or heat

Pregnant women are less tolerant of heat and may be more prone to fainting or heat stress. Although the risk is likely to reduce after birth, dehydration may impair breastfeeding. Exposure to prolonged heat at work, e.g. at furnaces or ovens, should be avoided. Maintenance or cleaning work in hot situations should also be avoided, particularly if this involves use of less secure forms of access such as ladders, where fainting could result in a serious fall.

The HSE do not consider that there are any specific problems from working in extreme cold although obviously appropriate precautions should be taken as for other workers, e.g. the provision of warm clothing.

Biological agents

3.14 Many biological agents in hazard groups 2, 3 and 4 (as categorised by the Advisory Committee on Dangerous Pathogens) can affect the unborn child should the mother be infected during pregnancy. Some agents can cause abortion of the foetus and others can cause physical or neurological damage. Infections may also be passed on to the child during or after birth, e.g. while breastfeeding.

Agents presenting risks to children include hepatitis B, HIV, herpes, TB, syphilis, chickenpox, typhoid, rubella (German measles), cytomegalovirus and chlamydia in sheep. Most women will be at no more risk from these agents at work than living within the community but the risks are likely to be higher in some work sectors, e.g. laboratories, health care, the emergency services and those working with animals or animal products.

Details of appropriate control measures for biological agents are contained in *Schedule 3* and Appendix 2 of the ACOP booklet to the *COSHH Regulations 2002*. There is a separate HSE publication on *Infection risks to new and expectant mothers in the workplace (1997)*. When carrying out a risk assessment in respect of new or expectant mothers, normal containment or hygiene measures may be considered sufficient, but there may be the need for special precautions such as use of vaccines. Where there is a high risk of exposure to a highly infectious biological agent it may be necessary to remove the worker entirely from the high risk environment.

Chemical agents

Substances labelled with certain risk phrases

3.15 The *Chemicals (Hazard Information and Packaging for Supply) Regulations 2002* (CHIP) require many types of hazardous substances to be labelled with specified risk phrases. Several of those are of relevance to new or expectant mothers:

- R40 – possible risk of irreversible effects;
- R45 – may cause cancer;
- R46 – may cause heritable genetic damage;
- R49 – may cause cancer by inhalation;
- R61 – may cause harm to the unborn child;
- R63 – possible risk of harm to the unborn child;
- R64 – may cause harm to breastfed babies.
- R68 – possible risk of irreversible effects.

The *CHIP Regulations 2002* are regularly subject to amendment and employers should be alert for other risk phrases which indicate risks in respect of new or expectant mothers.

Control of such substances at work is already required by the *COSHH Regulations 2002* or the separate Regulations governing lead and asbestos. Risk assessments in relation to pregnant women or those who have recently given birth may indicate that normal control measures are adequate to protect them also. However, additional precautions may be necessary, e.g. improved hygiene procedures, additional PPE or even restriction from work involving certain substances.

The ACOPs relating to the *COSHH Regulations* (contained in HSE booklet *L5*) together with *HS(G)193 COSHH essentials. Easy steps to control chemicals* provide further details of the types of precautions which may be appropriate.

Mercury and mercury derivatives

The HSE state that exposure to organic mercury compounds can slow the growth of the unborn baby, disrupt the nervous system and cause the mother to be poisoned. They consider there is no clear evidence of adverse effects on the foetus from mercury itself and inorganic mercury compounds. Mercury and its derivatives are subject to the *COSHH Regulations* and normal control measures (see above) may be adequate to protect new or expectant mothers (see HSE publications *L5* and *HSG 193*, as mentioned above; and *EH 17 Mercury and its inorganic divalent compounds (1996)*; and *MS 12 Mercury: medical guidance notes (1996)*).

Antimitotic (cytotoxic) drugs

These drugs (which may be inhaled or absorbed through the skin) can cause genetic damage to sperm and eggs and some can cause cancer. Those workers involved in preparation or administration of such drugs and disposal of chemical or human waste are at greatest risk, e.g. pharmacists, nurses and other health care workers. Antimitotic drugs can present a significant risk to those of either gender who are trying to conceive a child as well as to new or expectant mothers, and all those working with them should be made aware of the hazards.

The *COSHH Regulations* again apply to the control of these substances (see above). Since there is no known threshold limit for them, exposure must be reduced to as low a level as is reasonably practicable.

Agents absorbed through the skin

Various chemicals including some pesticides may be absorbed through the skin causing adverse effects. These substances are identified in the tables of occupational exposure limits in HSE booklet *EH 40 workplace exposure limits* (an updated version is published annually), in which some substances are accompanied by an annotation 'Sk'. Many such agents (particularly pesticides which are subject to the *Control of Pesticides Regulations 1986*) will also be identified by product labels.

Effective control of these chemicals is obviously important in respect of all employees (the *COSHH Regulations* applying once again) although risk assessments may reveal the need for additional precautions in respect of new or expectant mothers, e.g. modified handling methods, additional PPE or even their restriction from activities involving exposure to such substances.

The HSE booklet *L9 Safe use of pesticides for non-agricultural purposes (1995)* contains a further ACOP in relation to COSHH obligations.

Carbon Monoxide

Exposure of pregnant women to carbon monoxide can cause the foetus to be starved of oxygen with the level and the duration of exposure both being important factors. There are not felt to be any additional risks from carbon monoxide to mothers who have recently given birth or to breast-fed babies.

Once again, it is important to protect all members of the workforce from high levels of carbon monoxide (the *COSHH Regulations* require it) but it may also be necessary to ensure that pregnant women are not regularly exposed to carbon monoxide at lower levels, e.g. from use of gas-fired equipment or other processes or activities. The HSE Guidance Note *EH 43 Carbon monoxide (1998)* provides general guidance on carbon monoxide.

Lead and lead derivatives

High occupational exposure to lead has historically been linked with high incidence of spontaneous abortion, stillbirth and infertility. Decreases in the intellectual performance of children have more recently been attributed to exposure of their mothers to lead. Since lead can enter breast milk, there are potential risks to the child if breastfeeding mothers are exposed to lead.

All women of reproductive capacity are prohibited from working in many lead processing activities. All work involving exposure to lead is subject to the *Control of Lead at Work Regulations 2002*. Even where women of reproductive capacity are allowed to work with lead or its compounds, the blood-lead concentrations contained within the Regulations as an 'action level' and a 'suspension level' are set at half the figures for adult males. This is intended to ensure that women who may become pregnant already have low blood-lead levels. Once pregnancy is confirmed, the doctor carrying out medical surveillance (as required by *Regulation 10*) would normally be expected to suspend the woman from work involving significant exposure to lead. There is, however, no specific requirement in the Regulations themselves that this must happen. Detailed guidance on the Regulations is contained in HSE booklet L132 *Control of lead at work (2002)*.

Working conditions

3.16 The HSE guidance in *HS(G) 122* in relation to the 'working conditions' referred to in the *Management Regulations 1999, Regulation 16(1)* repeats the HSE's stated position in respect of work with DSE contained in Annex B to the HSE booklet *L26 Work with display screen equipment (2003)*. Their view is that radiation from DSE is well below the levels set out in international recommendations and that scientific studies taken as a whole do not demonstrate any link between this work and miscarriages or birth defects.

There is no need for pregnant women to cease working with DSE. However, the HSE recommend that, to avoid problems from stress or anxiety, women are given the opportunity to discuss any concerns with someone who is well informed on the subject.

Other aspects of pregnancy

3.17 Although the HSE booklet on new and expectant mothers at work (*HS(G)122*) draws attention to other features of pregnancy which employers may wish to take into account, it suggests that employers have no legal obligation to do so. However, this view has not been tested by the courts and an argument could be advanced that some of these represented 'working conditions' which involve risk to the mother or her baby within the terms of the *Management Regulations 1999, Regulation 16, Paragraph (1)(b)*.

The impact of these aspects will vary during the course of the pregnancy and employers are likely to need to keep the situation under review. A modified version of the appendix from the HSE booklet dealing with these 'other aspects' is provided below.

Aspects of pregnancy that may affect work

Aspects of pregnancy	*Factors in work*
Morning sickness	Early shift work Exposure to nauseating smells Availability for early meetings Difficulty in leaving job
Frequent visits to toilet	Difficulty in leaving job/site of work
Breastfeeding	Access to private area Use of secure, clean refrigerators to store milk
Tiredness	Overtime/long working hours Evening work More frequent breaks Private area to sit or lie down
Increasing size	Difficulty in using protective clothing Work in confined areas Manual handling difficulties Posture at DSE workstations (Dexterity, agility, co-ordination, reach and speed of movement may also be impaired)
Comfort	Problems working in confined or congested workspaces
Backache	Standing for extended periods Posture for some activities Manual handling
Travel – i.e. fatigue, stress, static posture	Reduce need for lengthy jouneys Change travel methods
Varicose veins	Standing/sitting for long periods
Haemorrhoids	Working in hot conditions/sitting
Vulnerability to passive smoking	Effective smoking policy (priority to the needs of non-smokers)
Vulnerability to stress	Modified duties Adjustments to working conditions or hours
Vulnerability to falls	Restrictions on work at heights or on potentially slippery surfaces
Vulnerability to violence (e.g. from contact with public)	Changes to workplace layout or staffing Modified duties
Lone working	Effective communication and supervision Possible need for improved emergency arrangements

A practical approach to assessing risks

3.18 Essentially risk assessment in respect of new or expectant mothers must be carried out by an employer if:

- there are women of childbearing age; and
- their work could involve risks to new or expectant mothers or their babies.

Most organisations employ women of childbearing age and, particularly in the case of large businesses, there may be a number of work activities that could create relevant risks. In some cases (e.g. those involving exposure to hazardous substances or radiation), it may be appropriate to stipulate that women are not allowed to work in certain activities, processes or departments once their pregnancy has been notified and/or for a finite period after they return to work after giving birth.

However, in many workplaces the issues will be far from clear cut – it would be neither necessary nor practical to prohibit pregnant workers from carrying out *any* manual handling or standing up or sitting down or climbing up ladders. Also it is difficult for any employer to identify in advance exactly what steps need to be taken to alter working conditions or hours of work so that risks can be avoided for every female employee who may become pregnant.

A much more practical approach is for the employer to develop a checklist similar to the sample provided at the end of this section. Such a checklist can be prepared by carrying out a review of the organisation's activities in order to identify risks which are present which *may* be of relevance in relation to new or expectant mothers. The risks described in the previous part of this chapter provide a good starting point from which a workplace-specific checklist can be developed. In some workplaces it may be appropriate to develop more than one such checklist (e.g. for production, maintenance and administration) because the profiles of risks are different.

Such a checklist should be completed by a suitable employer's representative (a personnel or health and safety specialist or the pregnant woman's manager perhaps) together with the pregnant woman herself. The checklist is intended to provoke discussion about the employee's possible exposure to the risks the employer has identified so that any necessary additional precautions (or changes to work practices) can be agreed. There is the opportunity to identify whether any further review of the situation is necessary. The process would be repeated once the mother returned to work after giving birth.

The sample checklist provided is for a firm of developers. Many of their female staff would only be exposed to risks in the office but some might visit actual or potential development sites where they may be exposed to different types of risks (difficult or dangerous access, contamination). Some female staff may be involved in work outside office hours and also travel extensively, sometimes to destinations abroad.

XYZ Development
New and expectant mothers at work

Risk assessment checklist

Name of employee:	**Work location:**	
Risk type	**Possible risk situations**	**Work practice changes/additional precautions agreed**
Manual handling	• Handling stationery or printed materials • Work in archives or stores • Moving office equipment, furniture, exhibition stands	
Posture and movement	• Unsatisfactory position at DSE workstation • Cramped working position • Standing for lengthy periods • Excessive use of stairs necessary • Access to difficult locations (e.g. ladder work) • Risks of slips or falls (e.g. on site) • Excessive travel (particularly by car or air)	
Physical and mental pressure	• Overall workload/deadlines • Difficult decision-making • Attendance at meetings at unsuitable hours • Dealing with aggressive or aggrieved persons	
Biological and chemical agents	• Contamination on sites/unusual workplaces • Work abroad (diseases, inoculations)	
Other issues	• Lack of toilet facilities/rest areas	

Signature (for XYZ Development): Date: Signature (new/expectant mother): Date for further review (if any):
(This form should be completed when the pregnancy is first notified and when the new mother returns to work.)

References

(All HSE publications)

3.19

1	HS(G) 165	Young people at work. A guide for employers (2000).
2	HS(G) 122	New and expectant mothers at work. A guide for employers (2002).
3	L121	Work with ionising radiation. *Ionising Radiations Regulations 1999* Approved code of practice and guidance (2000).
4	INDG 334	Working safely with radiation. Guidelines for expectant or breastfeeding mothers (2001).
5	L96	A guide to the *Work in Compressed Air Regulations 1996* (1996).
6	L103	Commercial diving projects offshore (1998).
7	L104	Commercial diving projects inland/ inshore (1998).
8	L105	Recreational diving projects (1998).
9	L106	Media diving projects (1998).
10	L107	Scientific and archaeological diving projects (1998).
11	L5	*Control of Substances Hazardous to Health Regulations 2002.* Approved codes of practice and guidance (2002).
12	—	Infection risks to new and expectant mothers in the workplace (1997).
13	HSG 193	COSHH Essentials. Easy steps to control chemicals.
14	EH 17	Mercury and its inorganic divalent compounds (1996).
15	MS 12	Mercury: medical guidance notes (1996).
16	EH 40	Workplace exposure limits (published annually).
17	L9	Safe use of pesticides for non-agricultural purposes (1995).
18	EH 43	Carbon monoxide (1998).
19	L132	Control of lead at work (2002).
20	L26	Work with display screen equipment (2003).

4 Carrying out risk assessments

In this chapter:

Planning and preparation	**4.1**
Who will carry out the assessments?	**4.2**
How will the assessments be organised?	**4.3**
Gathering documents	**4.4**
Consider who might be at risk	**4.5**
Identify the issues to be addressed	**4.6**
Variations in work activities	**4.7**
Checklist of possible risks	**4.8**
Making the risk assessment	**4.9**
Observation	**4.10**
Discussions	**4.11**
Tests	**4.12**
Further investigations	**4.13**
Notes	**4.14**
Assessment records	**4.15**
After the assessment	**4.16**
Review of the recommendations	**4.17**
Implementation of the action plan	**4.18**
Recommendation follow-up	**4.19**
Assessment review	**4.20**
References	**4.21**

Planning and preparation

4.1 Taking some time to plan and prepare for risk assessments should always pay dividends – both in saving time later on and in ensuring the assessment is more effective. The important aspects of this phase are set out below.

Who will carry out the assessments?

4.2 Decisions need to be taken about who will be involved in the risk assessment. In **2.6:** **WHO SHOULD CARRY OUT THE ASSESSMENT**, reference was made to the requirements of *Regulation* 7 of the *Management Regulations 1999* for employers to appoint competent health and safety assistance. Key attributes of competence are sufficient training, knowledge and experience together with other relevant qualities. Additional training may be necessary for some individuals, so that they are familiar with the purpose of risk assessments and are capable both of identifying risks and of evaluating the effectiveness of risk control measures.

As the HSE acknowledge in their guidance on risk assessment, in small businesses the employer or a senior manager may be quite capable of carrying out the risk assessment. Some organisations, both large and small, employ consultants to co-ordinate or carry out their risk assessment programme. Other employers create risk assessment teams who might be drawn from:

- managers;
- engineers and other specialists;
- supervisors or team leaders;
- health and safety specialists;
- safety representatives; and
- ordinary employees.

Whether the assessments are to be carried out by an individual or by a team, others will need to be involved during the process. Managers, supervisors, employees, specialists, etc. will all need to be talked to about the risks involved in their work and the precautions that are (or should be) taken.

A good combination is often to have an 'insider' and an 'outsider' working together. The 'insider' should be familiar with the work activity being assessed – the risks involved, the possible variations in the activity, the precautions available and their effectiveness and also whether what is being observed is typical. The 'outsider' is better able to question the status quo, to suggest possible risks that may have been overlooked, to put forward alternative precautions which may not have been considered and to bring to bear their knowledge and experience of similar situations elsewhere.

The 'outsider' role is often filled by a health and safety specialist (this is where employing an experienced consultant can be beneficial) although managers or safety representatives from other departments can also play the role effectively. The 'insider' is likely to be a manager, supervisor or safety representative from the department or section being assessed.

How will the assessments be organised?

4.3 In a small workplace it may be possible to carry out a risk assessment as a single exercise but in larger organisations it will usually be necessary to split the assessment up into manageable units. If done correctly, this should mean that assessment of each unit should not take an inordinate amount of time and also allows the selection of the people best able to assess an individual unit on a 'horses for courses' basis. Division of work activities into assessment units might be on the basis of:

- department or sections;
- buildings or rooms;
- parts of processes;
- product lines; and
- services provided.

As an illustration, a garage might be divided into:

- servicing and repair workshop;
- body repair shop;
- parts department;
- car sales and administration; and
- petrol and retail sales.

In dividing workplaces into units like this, it is important to take account of aspects which may be common to all assessment units, e.g. fire precautions, electrical supply, access roads, the impact of work activities on neighbours and others.

Where workplaces to be assessed have similarities to each other (e.g. groups of garages, retail chains or networks of offices), then use of the concept of a 'model risk assessment' may be appropriate. This is dealt with in more detail in **CHAPTER 6: MODEL RISK ASSESSMENTS**.

Gathering documents

4.4 There is a wide range of documents which might be of value during the risk assessment process. Some of these are likely to be internal documents whilst others will be reference material from the HSE and other external sources. Internal documents of relevance could include the following.

Previous risk assessments

There has been a legal requirement for risk assessment since the beginning of 1993. Even if previous assessments are considered to be inadequate or well out of date, they should be able to provide at least some useful information.

Specific risk assessments

Risk assessments carried out to comply with specific Regulations (e.g. COSHH or Noise) may also be useful. There may be an overlap with the more general risk assessment (e.g. eye protection against a corrosive substance may also protect against other risks) and the validity of the specific risk assessments may also be reviewed as part of the exercise.

Operating procedures

Good operating procedures should also incorporate health and safety considerations – the risks involved and the precautions required. Examples of this approach are provided in CHAPTER 5: ASSESSMENT RECORDS. Existing procedures may already do this to some extent and should be reviewed as part of the assessment process.

Safety handbooks, etc.

Many organisations produce safety handbooks or lists of safety rules in order to inform employees about risks and precautions. The awareness which should be created by such handbooks can be an important precaution in its own right and the detailed contents of the handbook can be useful in the risk assessment.

Training programmes and records

Checking the adequacy of employee training in health and safety should be an important part of risk assessment. The health and safety content of induction programmes and operator training should be reviewed, as should whether employees have actually received the training that they should have.

Accident/incident records

Work activities that have caused accidents or near misses in the past obviously involve potential risks which need to be evaluated. Accident and incident statistics and investigation report forms should be reviewed. Even in smaller workplaces without formal investigation systems, a study of records from the Accident Book is likely to be worthwhile.

Health and safety inspection reports

In organisations which have formal inspection systems, a review of the report forms is likely to reveal regular problems which are being identified or the occasional major problem with high risk potential (The need for regular health and safety inspections may also be one of the recommendations made as a result of the risk assessment.)

Amongst external documents which could be of relevance during risk assessments are as follows.

Regulations and approved codes of practice

Risk assessment involves an evaluation of compliance with legal requirements and therefore an awareness of the legislation applying to the workplace in question is essential. The HSE approved codes of practice (ACOPs) which accompany Regulations can be referred to in court and almost have the same legal status as the Regulations themselves – the onus is on accused persons or organisations to show that compliance with the law was achieved by other equally satisfactory means.

HSE guidance

The HSE publishes a wide range of booklets and leaflets providing guidance on health and safety topics. Some of these relate to the specific requirements of Regulations, others deal with specific types of risks whilst some publications deal with sectors of work activity.

All of these can be of considerable value in identifying which risks the HSE regard as significant and in providing benchmarks against which precautions can be measured. The guidance on specific types of workplaces (which includes engineering workshops, motor vehicle repair, warehousing, kitchens and food preparation, golf courses, horse-riding establishments and many others) should be an essential basis for those carrying out assessments in those sectors.

The HSE Books (P.O. Box 1999, Sudbury, Suffolk CO10 2WA, tel.: 01787 881165, fax: 01787 313995, website: www.hsebooks.co.uk) regularly publishes a detailed catalogue of HSE publications. Their booklet *Essentials of Health and Safety at Work* provides an excellent starting point for those in small businesses needing guidance or carrying out risk assessment. The booklet also contains a useful reference section to other HSE publications which may be of relevance.

HSE website

The main HSE website (www.hse.gov.uk) contains an extensive range of information that can be useful in carrying out risk assessments. The website provides access to portals which relate to particular industries such as agriculture, construction, engineering, food manufacture, haulage, health services, mining, motor vehicle repair, offshore oil and gas, railways, textiles and footwear. This list is steadily being extended to encompass a wider range of activities.

Information on specific health and safety topics can also be accessed using an index of subjects. The 'What's new' section of the website provides details of new HSE publications and consultative documents and HSE releases are also available. Many documents (including free HSE publications) can be downloaded from the website. Priced HSE publications can also be accessed and downloaded electronically via a subscription service, i.e. HSE Direct. Details of this are also available on the website.

Trade information

Many trade associations and similar organisations publish their own codes of practice and guidance on health and safety topics. Whilst these do not have the same legal significance as HSE publications, they nevertheless provide useful information on what others involved in similar work activities consider to be significant risks and appropriate precautions.

Information from manufacturers and suppliers

HSWA 1974, Section 6 places duties on manufacturers and suppliers to provide adequate information so that their products can be used in a safe manner. Such information might be in the form of a handbook associated with a piece of work equipment or a data sheet for a hazardous substance (relevant not only in a COSHH assessment but also in a general risk assessment if the substance is flammable or explosive). The information should identify the risks associated with the product and the precautions recommended by the manufacturer or supplier. It will be for those carrying out the risk assessment to determine what is appropriate in the circumstances of the product's use.

General reference books

There are many reference books which may be useful during the risk assessment process. Some, such as this handbook, provide guidance on how the risk assessment should be undertaken, whilst others provide technical advice on specific topics such as electrical safety, fire safety or machine guarding.

Consider who might be at risk

4.5 The risk assessment process must take account of all those who may be at risk from the work activities – both employees and others. It is important that all of these are identified.

Employees

Different categories of employees to take into account might include:

- production workers;
- maintenance workers;
- administrative staff;
- security officers;

- cleaners;
- delivery drivers;
- sales representatives;
- others working away from the premises; and
- temporary employees.

Some of these employees may merit special considerations:

- children and young persons (see **CHAPTER 3: SPECIAL CASES**);
- women of childbearing age, i.e. potential new or expectant mothers (see **CHAPTER 3: SPECIAL CASES**);
- employees with disabilities;
- people working alone;
- those working at night or weekends; and
- inexperienced staff.

Contractors and their staff

Most organisations utilise the services of contractors and their use has increased in recent years. The relationship between the host and the contractor will vary according to the nature of the services the contractor provides. These might include:

- construction or engineering projects;
- routine maintenance or repair;
- hire of plant and operators;
- support services, e.g. catering, cleaning, security, transport;
- professional services, e.g. architects, engineers, trainers; and
- supply of temporary staff.

Even though contractors have duties to carry out risk assessments in respect of their own staff, host organisations also have duties towards contractors' employees (as described in **CHAPTER 1: INTRODUCTION** and **CHAPTER 2: WHAT THE MANAGEMENT REGULATIONS REQUIRE**). Their duties will normally be greater in respect of temporary staff working alongside their own employees than for contractors providing specialist skills or services. However, even in the latter case, risks associated with facilities, equipment or materials supplied by the host, or created by the host's activities must be taken into account.

Others at risk

Other people that might be put at risk by the organisation's activities will depend upon the nature and location of those activities. Groups of people to be considered include:

- volunteer workers;
- co-occupants of premises;
- occupants of neighbouring premises;
- drivers making deliveries;
- visitors (both individuals and groups);
- passers-by;
- users of neighbouring roads;
- trespassers;
- customers or service users; and
- residents e.g. in the care or hospitality sectors.

Identify the issues to be addressed

4.6 The issues which will need to be addressed during the risk assessment process should be identified. This may be done in respect of the assessment overall or separately for each of the assessment units. Essentially this is a brainstorming process – based upon the knowledge and experience of those carrying out the assessment and the information gathered from the sources described earlier. In effect, this will create an initial list of headings and sub-headings for the eventual record of the risk assessment, although in practice this list is likely to be amended along the way. The sample assessment records contained in the next chapter will demonstrate how such a list can be built up for typical workplaces.

These headings are likely to consist of the more common types of risk, e.g. fire, vehicles, work at heights, together with some specialised types of risk associated with the work activities such as violence, lasers or working in remote locations. A checklist of possible risks to be considered during risk assessments is provided at **4.8: CHECKLIST OF POSSIBLE RISKS**, below. This includes some of the risks which have specific Regulations associated with them.

In some situations, particular notes may also be made to check on the effectiveness of the precautions which should be in place to control the risks, e.g. standards of machine guarding, compliance with PPE requirements or the quality and extent of training.

Variations in work activities

4.7 Consideration should also be given at this stage to possible variations in work activities which may create new risks or increase existing risks. Such variations might involve:

- fluctuations in production or workload demands;
- reallocation of staff to meet changing workloads;
- seasonal variations in work activities;
- abnormal weather conditions;
- alternative work practices forced by equipment breakdown/unavailability;
- urgent or 'one-off' repair work;
- work carried out in unusual locations; and
- the difference in work between days, nights or weekends.

Further variations may emerge later on in the assessment process.

Checklist of possible risks

4.8 This checklist is intended to assist in identifying which issues need to be addressed during risk assessments. In some cases, the headings and/or sub-headings might be used in the form shown, in other cases it may be more appropriate to combine them or modify the titles. (This is illustrated by practical examples in CHAPTER 5: ASSESSMENT RECORDS.)

Checklist of possible risks

Work equipment

- Process machinery
- Other machines
- Powered tools
- Hand tools
- Knives/blades
- Fork lift trucks
- Cranes
- Lifts
- Hoists
- Lifting equipment
- Vehicles

Services/power sources

- Electrical installation
- Compressed air
- Steam
- Hydraulics
- Other pressure systems
- Buried services
- Overhead services

Access

- Vehicle routes
- Rail traffic
- Pedestrian access
- Glazing

Work at height (see also CHAPTER 16)

- Ladders and stepladders
- Scaffolding
- Mobile elevating work platforms
- Falling objects

Work activity

- Burning or welding
- Entry into confined spaces
- Electrical work
- Excessive fatigue or stress
- Handling cash/valuables
- Use of compressed gases
- Molten metal

External factors

- Violence or aggression
- Robbery
- Large crowds
- Animals
- Clients' activities

Other factors

- Lasers
- Ultra violet/Infra red radiation
- Work-related upper limb disorder
- Stress

Storage

- Shelving and racking
- Stacking
- Silos and tanks
- Waste

Fire and explosion (see also CHAPTER 14)

- Flammable liquids*
- Flammable gases*
- Storage of flammables*
- Hot work

* DSEAR requirements

Work locations

- Heat
- Cold
- Severe weather
- Deep water
- Tides
- Remote locations
- Work alone
- Home working
- Poor hygiene
- Infestations
- Work abroad
- Work in domestic property
- Client's premises
- Site security
- Work-related driving

Risks/issues subject to separate assessment requirements

Hazardous substances (COSHH) (see **CHAPTER 9**)

Noise (see **CHAPTER 10**)

Manual handling (see **CHAPTER 11**)

Display screen equipment workstations (see **CHAPTER 12**)

PPE needs (see **CHAPTER 13**)

Fire and dangerous substances (see **CHAPTER 14**)

Asbestos in non-domestic premises (see **CHAPTER 15**)

Work at height (see **CHAPTER 16**)

Lead

Asbestos

Conformity of machinery

Major accident hazards (COMAH)

Ionising radiation

Vibration

Risks to (or from) others

Contractors
Volunteer workers
Co-occupants of premises
Neighbouring occupants
Delivery drivers
Visiting groups or individuals
Passers-by
Users of neighbouring roads
Trespassers
Customers or service users
Residents

Making the risk assessment

4.9 Good planning and preparation can reduce the time spent in actually making the risk assessment as well as enabling that time to be used much more productively. However, it is important that time is spent in work locations, seeing how work is actually carried out (as opposed to how it should be carried out).

Observation

4.10 Observation of the work location, work equipment and work practices is an essential part of the risk assessment process. Where there are known to be variations in work activities, a sufficient range of these should be observed to be able to form a judgement on the extent of the risks and the adequacy of precautions. Typical of evaluations which can be made through observations are:

- effectiveness of fixed guards;
- suitability and condition of access equipment;
- compliance with PPE requirements;
- observance of specified operating procedures or working practices;
- compliance with speed limits;
- standards of housekeeping and storage;
- acceptability of physical working conditions;
- suitability of manual handling practices;
- standards of lighting;
- suitability and availability of fire evacuation routes;
- locations and suitability of fire fighting equipment;
- availability of other emergency equipment;
- configurations of DSE workstations; and
- effectiveness of perimeter fencing.

These are only examples – the potential list is endless. However, by concentrating on the issues previously identified, it is possible to narrow down observations to those that are relevant for that area or that activity.

It should also be borne in mind that work practices may change once workers realise they are under observation. Initial or undetected observation of working practices may be the most revealing.

Discussions

4.11 Discussions with people carrying out work activities, their safety representatives and those responsible for supervising or managing them are also an essential part of risk assessment. Amongst aspects of the work that might be discussed are:

- possible variations in the work activities;
- problems that workers encounter;
- risks involved in the work, especially those that concern those carrying it out;
- workers' views of the effectiveness of the precautions available;
- the reasons some precautions are not utilised;
- suggestions for improving health and safety standards;
- the quality and manner of the training staff have received; and
- workpeople's awareness of emergency procedures.

These discussions should utilise open-ended questions such as:

- How do you get the large items onto the shelf?
- What happens if the fork lift truck breaks down?
- How do you carry out the task in high winds?
- What if no-one else is available to help?
- Why don't people adjust the guard correctly?
- How were you trained to carry out the task?
- How do you think the task could be done more safely?
- What do you do if the alarm goes off?

In evaluating the answers to such questions, it should always be borne in mind that those answering may well have different agendas to those carrying out the risk assessment.

Tests

4.12 In some situations, it may be appropriate to test the effectiveness of safety precautions. Examples of this are:

- the efficiency of interlocked guards or trip devices;
- stability of guard rails;
- security of locked enclosures;
- audibility of alarms or warning devices;
- suitability of access to remote workplaces, e.g. crane cabs, roofs; and
- whether lights or extraction fans operate when switched on.

In making any such tests, care must be taken by those carrying out the assessment not to endanger themselves or others, nor to disrupt normal activities. Suffering a fall as a result of an unstable hand rail or causing the premises to be evacuated through an injudicious test of an alarm button is not recommended!

Further investigations

4.13 Frequently, further investigations will need to be made before the assessment can be concluded. Such investigations may involve detailed checks on standards or records, or enquiries into how non-routine situations are dealt with. Examples of further checks or enquiries which might be appropriate are:

- detailed requirements of published standards, e.g. design of guards, thickness of glass;
- specifications of equipment or materials used, e.g. PPE design, substance data sheets;
- requirements contained in fire certificates;
- contents of operating procedures;
- maintenance or test records;
- contents of training programmes;
- training records;
- equipment not operating during the initial assessment;
- activities not in progress at the time of the assessment visit;
- contents of locked rooms or enclosures;
- intended responses to alarms;
- availability of equipment for non-routine activities; and
- experience of problems or malfunctions.

Once again, the potential list is endless although lines of further enquiry should be indicated by the observations and discussions during the initial phase of the assessment.

Notes

4.14 Rough notes should be made throughout the assessment process. It will be on these notes that the eventual assessment record will be based. It will seldom be possible to complete an assessment record 'on the run' during the assessment itself. The notes should relate to anything likely to be of relevance, such as:

- risks discounted as insignificant;
- more detail on risks known to be present;

- additional risks identified during the assessment;
- risks which are being controlled effectively;
- descriptions of precautions which are in place and effective;
- precautions which do not appear to be effective;
- alternative precautions which might be considered;
- problems identified or concerns expressed by others; and
- related procedures, records or other documents.

Such notes could be made on a clipboard or in a notebook, although some may be more comfortable using a hand-held computer or portable voice recorder. Subsequent analysis of the contents of a voice recorder to produce the assessment record may prove difficult however.

Assessment records

4.15 This phase of the risk assessment process concludes with the preparation of the assessment records. It may be preferable to record the findings in draft form initially, with a revised version being produced once the assessment has been reviewed more widely and/or recommended actions have been completed.

Decide on the record format

Many different formats are possible as can be seen in **CHAPTER 5: ASSESSMENT RECORDS**. Decisions need to be made on such questions as:

- Should separate records be prepared for different departments/sections/activities?
- Will a 'generic' assessment record need to be supported by 'dynamic' risk assessments in the field?
- Is a 'model' assessment record appropriate in this case?
- Should some or all of the assessment record be integrated into documented operating procedures?

Identify the section headings to be used

During the preparatory phase of the assessment, a list of risks and other issues to be addressed in each assessment unit was prepared as an aide-mémoire. This list will now need to be converted into section headings for the assessment records. As a result of the assessment, some of the headings may have been sub-divided into different headings whilst others may have been merged.

Prepare the assessment records

The rough notes taken during the assessment itself are extremely unlikely to be in a suitable order to link into the section headings to be used. The author has found it extremely useful to give each intended section heading a number and then work through all of the notes inserting the relevant section number (or numbers) alongside each note, usually using a different-coloured pen. A similar approach can be taken with computer-based records which could then be reshuffled together under their section headings.

Conversion of rough notes into formal assessment records can then take place more easily, although the process is still likely to be time consuming. Preparing the records will normally require at least half of the time which was spent in the workplace carrying out the assessment and sometimes might even take more time. It is wise not to allow too long to elapse between assessing in the workplace and preparation of the assessment record. Notes will seem much more intelligible and memory will often be able to 'colour in' between the notes.

Identify the recommendations

The assessment will almost inevitably result in recommendations for improvements and these must be identified. As can be seen in **CHAPTER 5: ASSESSMENT RECORDS**, some assessment record formats incorporate sections in which recommendations can be included, but in other cases separate lists will need to be prepared. The use of risk rating matrices (see **CHAPTER 7: SPECIALISED RISK ASSESSMENT TECHNIQUES**) may assist in the prioritisation of recommendations.

After the assessment

4.16 Whilst the recording of the risk assessment is an important legal requirement, it is even more important that the recommendations for improvement identified during the assessment are actually implemented. This is likely to involve several stages.

Review of the recommendations

4.17 In many organisations, the effective implementation of the recommendations is likely to involve others outside (and probably senior to) the risk assessment team. An appropriate group should be brought together to review the assessment findings and particularly the recommendations for improvements. The reasoning behind the recommendations can be explained and various alternative ways of controlling risks can be evaluated.

Changes to the assessment findings or recommendations may be made at this stage but the risk assessment team should not allow themselves to be browbeaten into making alterations that they do not feel can be justified. Similarly, they should not hold back on making recommendations they consider are necessary because they believe that senior management will not implement them. Once appointed to carry out risk assessments, the assessment team should carry out their duties to the best of their abilities in identifying what precautions are necessary in order to comply with the law – the responsibility for achieving compliance rests with their employer.

Some recommendations may need to be costed in respect of the capital expenditure or staff time required to implement them. It is unlikely that all recommendations will be able to be implemented immediately – there may be a significant lead time for the delivery of parts or the provision of specialist services from external sources. The risk assessment team should play an active part in costing and prioritisation issues.

Implementation of the action plan

4.18 Once costings and prioritisation have been agreed, the recommendations should be converted into an action plan with individuals clearly allocated responsibility for each element of the plan, within a defined timescale. Some members of the risk assessment team (particularly health and safety specialists) may have responsibility for implementing parts of the plan. In some cases, those given responsibilities may need guidance from the risk assessment team on the reasoning behind their recommendations or on the control measures the team believe are more likely to be successful.

Recommendation follow-up

4.19 Even in well-intentioned organisations, recommendations for improvement which have been fully justified and accepted are still not implemented. It is essential that the risk assessment process includes a follow-up of the recommendations made. Some of the recording formats provided in **CHAPTER 5: ASSESSMENT RECORDS** include reference to this.

As well as establishing that the improvements have actually been carried out, consideration should also be given to whether any unexpected risks have inadvertently been created, e.g. new eye protection may have reduced peripheral vision or a manual handling aid may have been introduced without staff being trained in how to use it correctly.

Once the follow-up has been carried out, the assessment record should be annotated or revised to take account of the changes made. In a minority of cases, a further follow-up may be appropriate to assess the position once changes have had time to 'bed in'.

If recommendations have not been implemented, there is a clear need for the situation to be referred back to senior management for them to take action to overcome whatever are the obstacles to progress. Particularly in larger organisations, the risk assessment team may prefer to do this via a more formal report.

Assessment review

4.20 Detailed reference was made in **CHAPTER 2: WHAT THE MANAGEMENT REGULATIONS REQUIRE** to the circumstances under which the *Management Regulations 1999* and the associated ACOP state that assessments must be reviewed. The ACOP also states that 'it is prudent to plan to review risk assessments at regular intervals'.

At this stage in the assessment process, it may be appropriate to determine how regular those intervals ought to be. The frequency should relate to the extent and the nature of the risks involved and the likelihood of creeping changes (as opposed to a major change which would automatically justify a review). Frequencies of review might, for example, be established as:

Construction or related activities	Every year
Activities of home care workers	Every two years
Industrial production processes	Every three years
Administrational activities	Every five years

It should be emphasised that these are suggested frequencies for review of the assessments. The review may conclude that the risks are unchanged, the precautions are still effective and that no revision of the assessments is necessary. The process of conducting regular reviews of assessments and, where appropriate, making revisions may be aided by the application of document control systems of the type used to achieve compliance with ISO 9000 and similar standards.

References

(HSE publications)

4.21

1 Essentials of health and safety at work (1994)

2 HSG 183 Five steps to risk assessment: Case studies (1998)

5 Assessment records

In this chapter:

Introduction	**5.1**
Contents of assessment records	**5.2**
Sample assessment format	**5.3**
Illustrative assessments	**5.4**
Estate agent's office	**5.4**
Supermarket	**5.4**
Motor vehicle repair workshop	**5.4**
Newspaper publisher	**5.4**
Out of school childcare group	**5.4**
Alternative assessment record formats	**5.5**
Employees' guides or handbooks	**5.6**
Contractors' manuals	**5.7**
Standard operating procedures	**5.8**

Introduction

5.1 *Regulation 3(6)* of the *Management Regulations 1999* states:

> Where the employer employs five or more employees, he shall record–
>
> (a) the significant findings of the assessment; and
>
> (b) any group of his employees identified by it as being especially at risk.

The accompanying ACOP refers to the record as representing 'an effective statement of hazards and risks which then leads management to take the relevant actions to protect health and safety'. It goes on to state that the record must be retrievable for use by management, safety representatives, other employee representatives or visiting inspectors. The need for linkages between the risk assessment, the record of health and safety arrangements (required by *Regulation 5* of the *Management Regulations 1999*) and the health and safety policy is also identified. The ACOP allows for assessment records to be kept electronically as an alternative to being in written form.

Contents of assessment records

5.2 The essential content of any risk assessment record should be:

- hazards or risks associated with the work activity;
- any employees identified as especially at risk;
- precautions which are (or should be) in place to control the risks (with comments on their effectiveness); and
- improvements identified as being necessary to comply with the law.

Other important details to include are:

- name of the employer;
- address of the work location or base;
- names and signatures of those carrying out the assessment;
- date of the assessment; and
- date for next review of the assessment.

(These might be provided as an introductory sheet.)

Sample assessment format

5.3 The sample risk assessment form provided contains specific sections for most of the contents referred to above. Groups of employees (or others) especially at risk can be included in the 'Risks identified' column, whilst the name and address of the employer could be overprinted on each sheet or provided at the front of a risk assessment file or manual. It includes a space for cross-references, e.g. to company procedures or training programmes, to other relevant risk assessments or to relevant HSE publications. The content of the form is not too dissimilar to that provided in the HSE's *Five steps to risk assessment* leaflet, but the layout is felt to be more user friendly.

Illustrative assessments

5.4 In the following pages of this chapter, the risk assessment methodology described in **CHAPTER 4: CARRYING OUT RISK ASSESSMENTS** is used to provide illustrations of how the process can be applied in a variety of workplaces or activities.

For each work situation:

- relevant risks or issues to be addressed are listed (utilising the checklist in **4.8: CHECKLIST OF POSSIBLE RISKS**);
- one of those risks or issues is selected;

- relevant Regulations, references and other key assessment points relating to that risk or issue are identified; and
- an illustration of how the completed assessment record might look is provided.

In some cases, the sample risk assessment format has been used, but in others the 'Recommendations' column has been omitted (any recommendations would be listed separately).

Later in the chapter, other possible formats are illustrated, demonstrating how risk assessments might be recorded in alternative documents (see **5.5: ALTERNATIVE ASSESSMENT RECORD FORMATS**).

Risk assessment		
Reference number:	**Risk topic/issue:**	Sheet ____ of ____
Cross references:		
Risks identified	**Precautions in place**	**Recommended improvements**
Signature(s)	**Name(s)**	**Date**
Dates for recommendation follow up		**Next routine review**

On the following pages you will find illustrative risk assessments for the following places of work:	
p. 78–79	Estate agent's office
p. 80–81	Supermarket
p. 82–83	Garage
p. 84–85	Newspaper publisher
p. 86–87	Out of school childcare group

Estate agent's office

Relevant risk topics are likely to include:

- *electrical equipment* – in the office and kitchen;
- *fire* (see **CHAPTER 14: FIRE AND DSEAR ASSESSMENTS**) – general risks only, no special risks;
- *access* – pedestrian access into and around the office;
- *manual handling* (see **CHAPTER 11: ASSESSMENT OF MANUAL HANDLING**) – of stationery, printed materials, records;
- *DSE* (see **CHAPTER 12: ASSESSMENT OF DSE WORKSTATIONS**) – used at workstations in the office;
- *hazardous substances* (see **CHAPTER 9: COSHH ASSESSMENTS**) – cleaning materials and office supplies;
- *asbestos* – possible presence of asbestos-containing materials in the premises.
- *lone working* – staff working alone in the office; and
- *personal safety outside the office* – visiting occupied and unoccupied property, with and without clients.

A sample assessment record for ***electrical equipment*** is provided opposite.

In conducting this risk assessment:

- the requirements of the *Electricity at Work Regulations 1989* must be complied with;
- the following HSE leaflets are likely to be relevant–
  - *INDG 236 Maintaining portable electrical equipment in offices and other low–risk environments*,
  - INDG 173 *Office wise* (this deals with office equipment and other office-related issues); and
- particular attention should be paid to–
  - the apparent condition of the electrical equipment,
  - inspection and maintenance arrangements,
  - the electrical isolator and distribution panel, and
  - any safety/guarding issues associated with the equipment.

Acorn Estate Agents, Newtown

Risk assessment		
Reference number: 1	Risk topic/issue: Electrical equipment	Sheet 1 of 1
Cross references: **HSE leaflets *INDG 236* and *173*, Risk assessment 3 (Access)**		
Risks identified	**Precautions in place**	**Recommended improvements**
Unsafe electrical equipment can present risk to staff and clients.		
Electrical supply system and fixed equipment (electrical heaters, water heater, socket outlets).	The isolator and distribution board (with earth leakage protection) are in the storeroom. The distribution board is clearly labelled. All electrical repair and maintenance work is carried out by a reliable, competent electrical contractor.	Ensure cleaning materials do not block access to the isolator or panel.
Mobile and portable equipment including fridge, photocopier, computer equipment, shredder, fans.	Staff are encouraged to report immediately any defects in sockets, plugs, cables, connections, etc. Electrical equipment is checked during quarterly office health and safety inspections. Equipment is inspected and/or tested by electrical contractors every two years. Photocopier maintained under contract.	Some personal equipment (e.g. kettles, radio) is in use. This should be forbidden unless the equipment has been tested and/or inspected by a competent person.
The photocopier, shredder and computer printers contain in-running nips and other dangerous parts.	Access to these dangerous parts prevented by a combination of fixed guards and interlocked guards.	Staff should be warned about: – hot drums when cleaning photocopier blockages; – risks to tie wearers when using the shredder.
Signature(s) K Stephenson, R Lewis	**Name(s)** K Stephenson, R Lewis	**Date** 4/11/98
Dates for recommendation follow up February 1999	**Next routine review** November 2003	

Supermarket

Relevant risk topics are likely to include:

- *food processing machinery* – in the delicatessen;
- *knives* – delicatessen and butchery;
- *fork lift trucks* – warehousing areas;
- *vehicle traffic* – delivery vehicles, customer and staff vehicles;
- *vehicle unloading* – in the goods inward area;
- *pedestrian access* – external to and inside the store;
- *glazed areas* – store windows, display cabinets;
- *electrical installation* – the power system within the store;
- *electrical equipment* – including tills, cleaning equipment, office equipment;
- *shelving* – in the store;
- *racking* – in warehouse areas;
- *refrigerators* – in the store and warehouse;
- *fire* – general risks only, precautions must take account of customers;
- *aggression* – e.g. from unhappy customers;
- *possible robbery* – the supermarket will hold large quantities of cash;
- *WRULD* – for checkout operators;
- *hazardous substances* – cleaning materials, office supplies;
- *manual handling* – e.g. shelf stacking, movement of trolleys;
- *DSE* – used at workstations in the office;
- *PPE requirements* – throughout the supermarket;
- *asbestos* – possible presence within the premises; and
- *use of contractors* – for maintenance and repair work.

A sample risk assessment record for ***vehicle traffic*** is provided opposite.

In conducting this risk assessment:

- the requirements of the *Workplace (Health, Safety and Welfare) Regulations 1992* must be complied with;
- HSE booklet *HS(G) 136 Workplace Transport Safety* is likely to be relevant; and
- particular attention should be paid to
  - signage and road markings,
  - observation of vehicle movements and speeds, and
  - conditions during busy periods and hours of darkness.

ACME Supermarkets, Oldborough

Risk assessment		
Reference number: 4	Risk topic/issue: Vehicle traffic	Sheet 1 of 1
Cross references: HSE booklet *INDG 136*, Risk assessments 5 (vehicle unloading) and 6 (pedestrian access)		
Risks identified	**Precautions in place**	**Recommended improvements**
Vehicles making deliveries to the 'Goods inward' bay present risks to each other and to any pedestrians on the access road.	Prominent 10 mph signs are in place on the roadway. The speed limit is enforced effectively. Signs prohibit use of the road by pedestrians and customer or staff vehicles.	
There are also risks to other road users as they leave and rejoin the main road.	There are give way signs and road markings at the junction with the main road. Vehicles are parked in a holding area prior to backing up to the 'Goods Inward' bay.	Reposition the advertising sign which partly blocks visibility on rejoining the main road.
	Movement of vehicles is controlled by a designated member of the supermarket staff.	Provide this designated staff member with a high visibility waterproof jacket.
	The area is well lit by roadside lamps and floodlights on the side of the building.	
Customer and staff vehicles circulating in the car park area present risks to each other and to pedestrians in the area.	The access road around the car park is one way and well indicated by signs. The entrance and exit are well separated from each other and the goods access road.	
There are also risks to other road users at the entrance from and access back into the main road.	There are prominent 15 mph signs around the roadways (some vehicles exceed this speed). There are give way markings and signs at all roadway junctions.	Provide clearly marked speed ramps at suitable locations.
	Parking bays are well marked. Pedestrian crossing points are clearly marked and signed.	The surface of some bays in the southwest corner of the car park should be repaired.
	The area is well lit by lighting towers which are protected at the base. Security staff inspect and, where necessary, salt roadways in icy or snowy weather.	Include the goods access road, Goods inward bay and car park in routine safety inspections.
Signature(s) T Burke, G Hare **Dates for recommendation follow up** January 1999	**Name(s)** T Burke, G Hare **Next routine review** October 2001	**Date** 12/10/98

Motor vehicle repair workshop

Relevant risk topics are likely to include:

- *tools and equipment* – bench machinery, portable powered tools, hand tools;
- *vehicle hoists* – extending to include engine or gearbox lifters, if appropriate;
- *access* – for vehicles and pedestrians;
- *burning and welding* – storage and use of oxyacetylene equipment;
- *electrical installation* – the power system within the workshop;
- *compressed air* – the compressor and air receiver;
- shelving, racking, etc. – storage of parts and tools;
- *oil tanks* – inside and outside the workshop;
- *fire* – precautions within the workshop, including control of flammable liquids;
- *vehicle recovery work* – if this is applicable;
- *hazardous substances* – engine fumes, valeting work, battery acid, parts cleaning, etc.;
- *manual handling* – of parts and materials; and
- *PPE requirements* – in the workshop and parts department (if separate).

(This list does not take account of risks associated with body repair work, the parts department, petrol and other sales, car sales showrooms or administrative areas).

A sample assessment record for ***tools and equipment*** is provided opposite.

In conducting this risk assessment:

- the requirements of the *Provision and Use of Work Equipment Regulations 1998* (PUWER) and the *Electricity at Work Regulations 1989* must be complied with;
- HSE booklet *HS(G)67 Health and safety in motor vehicle repair* is likely to be relevant; and
- particular attention should be paid to–
  - the power sources used for portable equipment;
  - potential for trailing cables and hoses;
  - physical condition of equipment and its manner of use; and
  - inspection and maintenance arrangements.

Goodfellows Garage, Tupton

Risk assessment		
Reference number: 1	Risk topic/issue: Tools and equipment	Sheet 1 of 1
Cross references: HSE Booklet HS(G)67 Risk assessments 3 (access), 5 (electrical installation) and 6 (compressed air)		
Risks identified	**Precautions in place**	**Recommended improvements**
The equipment below could represent a risk to any member of the workshop staff.		
30 tonne hydraulic press.	No guards. Slow moving. Apprentices directly supervised until competent to use alone.	Re-secure the floor mountings which have worked loose.
Bench-mounted drilling machine.	Adjustable chuck guard in place.	
Bench-mounted grinder/wire brush.	Standard partial enclosing guards for both.	
	Tool rest for grinding wheel. Spindle speed clearly marked. Technician trained to mount grinding wheels. Goggles available on shelf above bench.	Tool rest requires adjusting closer to the wheel.
Portable drilling machines.	Air-powered or battery-operated machines. All technicians provided with safety spectacles.	
Portable hand lamps.	Low voltage, totally enclosed type. Compressed air and electric power from suspended supply unit at each servicing bay (minimising tripping hazards).	Replace lamp found with cracked enclosure.
Hand tools (both supplied by the company and belonging to technicians).	Staff are expected to check the condition of all equipment before use. (All equipment was in satisfactory condition apart from the exceptions noted). The manager inspects the workshop and the equipment in it every three months. All electrical equipment is inspected and tested (where appropriate) annually by a competent electrical contractor.	Remind technicians that defective equipment must not be used, quoting the above examples.
Signature(s) A Goodfellow **Dates for recommendation follow up** March 2000	**Name(s)** A Goodfellow	**Date** 2/2/2000 **Next routine review** February 2003

Newspaper publisher

A large workplace like this would need to be divided into assessment units (see **CHAPTER 4: CARRYING OUT RISK ASSESSMENTS**), which might consist of:

- *common facilities, services, etc.* – e.g. fire, electrical supply, lifts, vehicle traffic, presence of asbestos;
- *reel handling and stands* – supply of reels of newsprint to the press area;
- *platemaking* – equipment and chemicals used to produce printing plates;
- *printing press* – press machinery, solvents, noise, etc.;
- *despatch* – inserting equipment, newspaper stacking, strapping and loading;
- *circulation and transport* – distribution and other vehicles, fuel, waste disposal;
- *maintenance* – workshops, garage, maintenance activities; and
- *offices* – editorial and administrative areas.

Selecting the *reel handling and stands* unit, risk topics are likely to include:

- *fork lift trucks* – used to unload, transport and stack reels;
- *reel storage* – stability of stacks, access issues;
- *reel handling equipment* – hoists, conveyors and the reel stands feeding the press;
- *wrappings and waste* – removal of wrappings, storage and disposal of waste paper, etc.; and
- *noise and dust* – from the operation of the reel stands and nearby press.

A sample assessment record for ***reel handling equipment*** is provided opposite.

In conducting this risk assessment:

- the requirements of the PUWER 1998, the *Lifting Operation and Lifting Equipment Regulations 1998* and the *Manual Handling Operations Regulations 1992* must be complied with;
- reference may need to be made to relevant BS specifications and to other standards for conveyors or specialist handling equipment; and
- particular attention should be paid to
  - guarding standards and the possible presence of unguarded dangerous parts,
  - any need for manual handling of the reels,
  - training issues relating to the above, and
  - statutory examination records (for the hoist).

Grimtown News

Risk assessment		
Reference number: B3	Risk topic/issue: Reel handling and stands – reel handling equipment	Sheet 1 of 1
Cross references: Risk assessments B1 (forklift trucks), B2 (reel storage), B5 (Noise and dust)		
Risks identified	**Precautions in place**	**Recommended improvements**
The equipment below presents risks to all staff working in the area.		
Reel hoist – Carries reel down from the reels store to the reel stand basement. It is fed by forklift trucks and feeds onto the roller conveyor system.	Slow moving hoist protected by a substantial mesh guard. No need for access within the hoist enclosure. Sign states 'Do not ride on hoist'. Gap between base of hoist and roller conveyor (no shear trap). Statutory examinations by Insurance Engineers (kept by Works Engineer).	Replace this sign by one complying with the *Safety Signs Regulations*.
Roller conveyor – This is in a T formation and consists of powered and free-running rollers with a turntable at the junction. Reels are transferred to holding bays or floor-based trolleys by fork lift truck.	All drives for the conveyor are fully enclosed and there are no in-running nips. The conveyor is protected by kerbs from fork lift damage. Signs prohibit climbing on the conveyor system.	Replace these signs as above.
Floor-based trolleys – This system carries reels right up to the transfer carriages which load them onto the reel stands.	Reels can be easily moved by one person using the trolley system. Reels can safely be rolled across the floor onto the trolleys if the correct technique is used.	Include formal training on correct techniques in the induction programme for new staff in the area.
Some reels must be transferred manually from the holding bays onto the trolleys.	Two persons are required to move the transfer carriages up to the reel stands.	
Reel stands – The reel stands rotate mechanically ensuring a constant web feed to the press.	There are no accessible in-running nips. Operatives do not need to approach the rotating reels.	
The rotating drive shaft for the reel stands is approximately 4 metres above ground level.	This shaft is considered to be 'safe by position' for normal operating purposes.	Ensure a safe system of work for maintenance work near the shaft, e.g. by using a permit to work procedure.
Signature(s) M Betts S Brown **Dates for recommendation follow up** June 2000	**Name(s)** M Betts, S Brown	**Date** 26/3/00 **Next routine review** March 2002

Out of school childcare group

Such a group would need to assess risks to its staff and to children in its care, both on its own premises and in outside activities. Relevant risk topics are likely to include:

- *equipment* – cleaning, kitchen equipment, TVs and videos, play equipment;
- *premises and facilities* – internal and external access, electrical installation;
- *control of children* – supervision, permitted play areas, attendance records;
- *fire* – precautions and emergency procedures;
- *external activities* – walks, local visits, special excursions;
- *personal safety* – controlling access to premises, suspicious prowlers, aggressive parents;
- *hazardous substances* – cleaning materials;
- *manual handling* – movement of furniture and equipment; and
- *accidents, etc.* – dealing with accidents, illness, infectious diseases.

A sample assessment record for ***external activities*** is provided opposite.

The record only refers to the risks identified and the *expected* precautions. In effect, staff will need to carry out a dynamic risk assessment prior to any external activities taking place. This is likely to be more formal in the case of special excursions. Any improvements recommended (e.g. a checklist for special excursions) would be referred to in a separate document.

In conducting this risk assessment:

- there are no specific Regulations applying, although note must be taken of general obligations to safeguard staff and children;
- there is no specific HSE guidance (although guidance from education, local authority or childcare partnership sources may be available); and
- particular attention should be paid to
  - observing at least one external activity taking place,
  - discussing potential problems and solutions with staff.

Funkids, Beachville

<table>
<tr><th colspan="3">Risk assessment</th></tr>
<tr><td>Reference number: 5</td><td>Risk topic/issue: External activities</td><td>Sheet 1 of 1</td></tr>
<tr><td colspan="3">Cross references: Risks assessments 6 (personal safety) and 3 (control of children)</td></tr>
<tr><td>Risks identified</td><td colspan="2">Expected precautions</td></tr>
<tr><td>General
Any external visit of activity can create additional or different types of risks to children, whether by putting them closer to risk situations (e.g. roads or water) or increasing the difficulties in controlling them.</td><td colspan="2">External visits and activities have many benefits both for children and staff.
Risks can be minimised with a little thought and planning.
Staffing levels must be considered – at least two staff must go with the children making the visit and at least two must stay if any children are to remain on Club premises. The numbers and capabilities of the children making the visit must be taken into account.
A mobile telephone should normally be taken by the staff member in charge of the external activity.</td></tr>
<tr><td>Local visits
Local visits may be made to:
– parks, woodland or play areas;
– the beach;
– sports centres or swimming pools;
– shops.</td><td colspan="2">Activities must be chosen that relate to the ages and capabilities of the children involved. Some activities may require the children to be told to bring appropriate clothing or equipment with them, e.g. towels and swimming costumes.
Weather conditions must be assessed before the visit is allowed to go ahead.
Travel routes should be planned to utilise footpaths as much as possible and to avoid road crossings, especially those on busy roads without controlled crossing points.
Consideration may need to be given to the need for wheelchair access.
Rules for walking should be clearly established and communicated to the children. These will normally involve a staff member at the front and the rear of the group.
If public transport is to be used, care must be taken to avoid the group being split up.</td></tr>
<tr><td>Special excursions
Occasional visits may be made to major attractions or facilities which are further afield.</td><td colspan="2">Locations and activities must be chosen that relate to the ages and capabilities of the children involved.
Staffing levels should normally be at least at a level of 1:4.
Volunteer assistance, e.g. from parents may be necessary.
Transport must be booked from a reputable source (public transport should be avoided if possible). Coaches or minibuses must have seatbelts suitable for children and arrangements for wheelchair access may be necessary.
Refreshment arrangements must be planned in advance – either by requiring the children to bring food and drink with them or by pre-booking refreshments at the visit location.
Children must be required to bring suitable bad weather clothing if the visit involves significant time outdoors.
Parents should be advised in writing of when the group is expected to return.</td></tr>
<tr><td>Signature(s) C Sands A Castle
Date for next routine review April 2002</td><td>Name(s) C Sands, A Castle</td><td>Date 27/4/00</td></tr>
</table>

Alternative assessment record formats

5.5 Providing it contains all the essential ingredients, a risk assessment can be recorded in alternative formats to those used so far in this chapter. The remaining illustrations utilise:

- an employees' guide;
- a contractors' manual; and
- a standard operating procedure.

Employees' guides or handbooks

5.6 This format can be particularly useful when staff are working away from their employer's base. The example opposite is adapted from a guide prepared for engineers involved in *installing and maintaining specialist garage equipment* such as vehicle hoists, lubrication systems, vehicle washes, etc. The topics covered include:

- legal obligations;
- liaison with the customer;
- general access;
- work at height;
- manual handling;
- work equipment;
- use of lifting equipment;
- electrical work;
- use of gases;
- fire safety (including DSEAR issues);
- hazardous substances;
- asbestos;
- personal protective equipment;
- dealing with accidents and first aid.

The example illustrated is the section dealing with *fire safety*. As can be seen, although this is a generic risk assessment providing guidance to engineers, they will still need to carry out a dynamic risk assessment before starting work at a location.

Garage equipment engineers' health and safety guide

Section 10 Fire safety

10.1 General risks

Staff are exposed to a general risk of fire in any premises that they work in. They should make themselves aware of key elements of fire precautions, particularly:

- the location of fire alarm break glass call points (which must be activated if they become aware of a fire);
- the fire assembly point, at which they must assemble in the event of fire.

When working in more complex premises it may be necessary to ensure that they are aware of relevant fire evacuation routes.

10.2 Higher risk activities

Risks will be much greater when work has to be carried out near flammable liquids or gases or in areas where explosive atmospheres may be present. This will be particularly the case where hot work (e.g. burning or welding) is involved. Risks are likely to be highest in the vicinity of:

- vehicle finishing units;
- paint mixing areas;
- flammable liquid stores;
- oil storage facilities.

The precautions necessary should be discussed with the customer and are likely to involve some or all of the following:

- suspension of paint spraying or mixing activities;
- removal of flammable liquid containers and other combustible materials from affected areas (or their covering with non-flammable sheets or protective screens);
- use of suitably protected electrical equipment in potentially explosive areas;
- ensuring suitable fire extinguishers are available;
- provision of a 'fire watcher' by the customer.

For work in positions where access and egress is difficult, consideration may need to be given as to how staff will escape from such positions in the event of fire.

Contractors' manuals

5.7 Many manufacturers and suppliers of equipment are conscious of the importance of ensuring that sub-contractors who install their products do so with due regard to health and safety. This concern is not purely altruistic but is partly borne out of pressure from clients and/or principal contractors who are mindful of their own obligations under the *CDM Regulations 1994*. Installation contractors are often only small companies without risk assessment documentation of their own – some are too small for recording of risk assessments to be a legal requirement.

The demands for risk assessment documentation from clients or principal contractors can quite legitimately be met by the manufacturer or supplier preparing a generic risk assessment relating to the installation of their product in the form of an installation contractors' manual or guide. However, such a manual would need to be supported by some sort of an assessment relating to the requirements of a specific project – as referred to in the illustration earlier in the chapter relating to the installation of burglar alarms and closed circuit TV. There would also need to be a monitoring system in place to ensure that contractors complied with the requirements of the manual, probably through a system of spot checks by the manufacturer or supplier.

The example opposite is taken from such a manual relating to the *installation of communication and alarm equipment inside residential property*, the majority of which would be occupied during installation. The topics covered in the manual included:

- *residents and the public* – how they should be protected;
- *services* – electricity, telecommunications, gas, water, sewage;
- *asbestos* – likely to be present in older property;
- *personal protective equipment* – requirements for sites generally and for specific activities;
- *electrical work* – when working on or connecting into existing supplies;
- *work equipment* – use of power tools;
- *work at heights* – in stairways, lobbies, lofts;
- *fire* – general precautions;
- *hazardous substances* – sealants, adhesives, electrical cleaners;
- *manual handling* – particularly movement of heavy control panels; and
- *first aid and accidents.*

The illustrative example provided has been slightly adapted from the section relating to *residents and the public.*

Installation contractors' manual

Section 1 Residents and the public

Possible Risks

Installation work is often carried out within buildings where residents are still present. Other members of the public may visit the buildings and pass through or close to the installation site.

The nature of our products means that residents are often elderly, frail and lacking in mobility.

Visitors may also include young children who will be unaware of risks that might be obvious to adults and such children may also be unable to understand warning signs.

Precautions:

- Prior to starting work, there should be communication with affected residents about the nature of the work, its likely duration and any impact on them – particularly in respect of access.
- Every effort should be made to maintain clear pedestrian access through application of good housekeeping standards and careful choice of storage locations.
- Vehicles should be parked sensibly, leaving access clear for residents and other vehicles, particularly those used by the emergency services.
- At some locations it may be necessary to bring equipment onto site using a vehicle and then park the vehicle off site.
- Where special access risks are created, e.g. by excavations in roadways or footpaths or blockages of staircases or corridors, use of barriers and warning signs should be made.
- Every effort should be made to avoid leaving uncovered excavations or blocked access routes for extended periods, particularly overnight and at weekends.
- Staff carrying out installation work must keep close control over the equipment and materials they are using, particularly if these might be dangerous to children, etc.

Standard operating procedures

5.8 In work activities, there is often a tendency to place issues such as production, quality of product and health and safety into separate compartments, each covered by separate procedures or rules. This often makes life difficult at shop floor level where a single integrated standard operating procedure is not only much less confusing but also can be an invaluable training aid.

A risk assessment approach should be taken in the development of such procedures, with the risks and the appropriate precautions being identified at each stage of the manufacturing process. The example opposite has been extracted and adapted from a procedure for the *manufacture of pre-stressed concrete beams*.

Whilst the existence of a procedure of this type would not totally eliminate the need for other risk assessments, the process was so much at the heart of this small company's operations that the amount of other risk assessment documentation was greatly reduced. Risk assessments similar to those described earlier in the chapter were prepared for (see **5.4: ILLUSTRATIVE ASSESSMENTS**):

- *fork lift trucks and lifting equipment* – operator training, maintenance, statutory examinations;
- *beam stacking and storage* – general rules and precautions;
- *hazardous substances* – extremely limited use;
- *fire* – a low risk in this environment;
- *noise* – from saws and other process equipment;
- *work equipment* – guarding of saws, fitting of cutting discs, operator training; and
- *PPE requirements* – including more detail on specifications of PPE types.

The procedure emphasises the precautions which must be taken at each stage in the process without necessarily going into the same level of detail on the risks involved. However, in many cases the risks are fairly self-evident, especially for those familiar with industrial equipment such as Stihl saws. In some situations, though, it is important to emphasise risks which may be less apparent, e.g. the risk of stack collapse. Operatives are more likely to take precautions if they understand the reasons why such precautions are necessary.

Manufacturing and health and safety procedure

Activity		Key health and safety points
2	**Removal of cast beams**	
2.1	Attach the appropriate lifting beam to the forks of the fork lift truck.	The fork lift truck must only be driven by licensed drivers.
2.2	Locate the hooks of the lifting beam into the lifting eyes cast into the concrete beams.	
2.3	Raise the lifting beam a few centimetres using the fork lift truck.	The fork lift truck driver should beware of excessive resistance. It may be necessary to take remedial action if beams are sticking in the moulds. (The supervisor should be consulted in such cases.)
2.4	An operative must then stand on the moulds and cut out the spacers and wires using a Stihl saw or electric saw. Spacers must be placed in the 'wheelie bin' provided.	Only trained operatives may use the Stihl saw or electric saw. The saw guard must be in position. Care must be taken standing on the moulds and using the saw. The operative must wear: • eye protection (a face visor or goggles); • safety footwear; • hearing protection (muffs or plugs); • a dust mask; and • suitable gloves. (others working in the immediate vicinity must wear similar protection)
2.5	Lift out the row of beams using the fork lift truck.	
2.6	Place the beams on the designated stack.	Care must be taken to ensure that the stack is stable. No stacks must be more than ten beams high.
2.7	At the same time, an operative at the rear of the stack places wooden spacers below the beams and removes the lifting beam hooks from the lifting eyes.	This operative must remain alert for possible stack collapse. Care must be taken in positioning his hand to avoid trapping. Suitable gloves and safety footwear must be worn.

6 Model risk assessments

In this chapter:

Introduction	**6.1**
Implementation and adaptation	**6.2**
Sample model assessments	**6.3**

Introduction

6.1 The concept of a model risk assessment is referred to in the ACOP to the *Management Regulations 1999* which states 'Employers who control a number of similar workplaces containing similar activities may produce a "model" risk assessment reflecting the core hazards and risks associated with these activities.' The ACOP also refers to their development by trade associations, employers' bodies or other organisations concerned with a particular activity.

This chapter contains examples of model risk assessments used in office environments and in the motor vehicle trade, and the author has also seen the technique used successfully in the retail sector and in residential care homes.

Implementation and adaptation

6.2 The issue of a model risk assessment by a major employer or trade association will achieve very little if its findings are not implemented at local level and the assessment adapted to match local circumstances. The ACOP states that employers or managers at each workplace must:

- satisfy themselves that the 'model' is appropriate to their type of work; and
- adapt the 'model' to the detail of their own actual work situations, including any extension necessary to cover hazards and risks not referred to in the 'model'.

The examples provided in this chapter all contain a column-headed 'local arrangements' in which each location is expected to provide details of how they implement the precautions identified as being necessary. This approach works particularly well when used in conjunction with an auditing programme. For example, the auditor can identify from the document who has been made responsible for maintenance and testing of the fire alarm and therefore knows whom to approach to examine the relevant records. The records may demonstrate full compliance with the required standards or alternatively that precautions fall a long way short of those intended.

The existence of such an auditing programme should be an integral part of the 'management cycle' required by the *Management Regulations 1999, Regulation 5* (as described in more detail in **CHAPTER 8: IMPLEMENTATION OF PRECAUTIONS**). The audit constitutes part of the 'monitor' stage of the management cycle. This and the 'review' stage of the cycle can be particularly embarrassing for local managers who have simply placed the model risk assessment on a shelf.

Sample model assessments

6.3 The remainder of this chapter provides examples of work situations where model risk assessments have been developed successfully, together with sample extracts from those model assessments.

Offices

The examples provided on the following pages are modified extracts from a model risk assessment developed for the offices occupied by a regional development agency.

Topics covered by the assessment included:

- fire;
- accidents and other emergencies – including first aid arrangements, guidance on bomb threats;
- visitors and contractors – access control, contractor selection and management;
- access – inside and outside the offices;
- office and kitchen equipment – including lifts;
- services;
- hazardous substances – predominantly office supplies and cleaning materials (assessment of asbestos-containing materials should be dealt with separately);
- manual handling – stationery, archived records, etc; and
- DSE – arrangements for eye tests and workstation assessments.
- personal safety outside the office – including personal attack, road travel, higher-risk sites and PPE guidance;

- home working – an increasing practice in many organisations; and
- work abroad – medical issues and risks which may be greater in foreign countries.

The first example contains the subsections dealing with *fire alarms and emergency lighting* (other subsections deal with general fire matters, fire evacuation routes, evacuation procedures, evacuation drills, fire fighting equipment and fire prevention. Forms for recording tests, drills, etc. were provided at the end of the section).

The second example is the section relating to *services*. As can be seen, it is primarily concerned with arrangements for maintenance and repairs. The agency has established procedures for regular health and safety inspections which should be able to identify where repairs are necessary. Some offices, e.g. those with electrical heating would identify some subsections as 'not applicable'.

1 Fire

Details of risks	Expected precautions	Local arrangements
	1.3 FIRE DETECTION AND ALARMS	
A fire could break out at any time within the Agency's own offices, elsewhere in shared buildings or within adjacent property.	Arrangements must be in place to ensure that electrical fire detection and alarm systems are checked by specialist companies at least once per year.	Alarm and detectors maintained by:
The Agency's own staff would be at risk, together with any visitors or contractors present on the premises.	The fire alarm system must be tested at least weekly, using different call points in rotation. Records must be kept of the date of the test, the call point(s) used and the condition of the alarm.	Alarm tested and records kept by:
Agency activities would not normally create any abnormal fire risk but work by contractors may do – see *Section 3* of the risk assessment.	(A suitable record form is provided at the end of *Section 1*, although a separate Fire log book may be used, if preferred.)	
	1.4 EMERGENCY LIGHTING	
(This general description of risks is contained in the first part of this section).	Where necessary, offices will be provided with emergency lighting. (This is normally required for internal staircases, corridors, etc.) Arrangements must be in place to ensure that lighting is checked by a specialist company at least once per year.	Emergency lighting maintained by:
	The emergency lighting must be tested at least every six months. Records of the date of each test must be kept which include the date and details of any faults found. (A suitable record form is provided at the end of *Section 1*.)	Emergency lighting tested and records kept by:

6 Services

Details of risks	Expected precautions	Local arrangements
The fixed electrical installation, any gas supplies and any central heating or hot water boilers all present possible risks to Agency staff and others on the premises.	*6.1 FIXED ELECTRICAL INSTALLATION* The fixed electrical installation (distribution boards, isolators, circuit-breakers or fuses, cables and conduits and socket outlets) in each office must be inspected by a competent person (e.g. an electrician) *at least every 5 years*.	Inspection of the electrical installation was last carried out by: on:
	Distribution boards, isolators, circuit-breakers and fuses should be labelled as appropriate.	
	Access to isolator switches should be kept readily available. (Where it is important to keep isolators secure, keys to locked rooms or cupboards should be kept in glass-fronted boxes or otherwise available for emergency access.)	
	6.2 GAS INSTALLATION AND EQUIPMENT Locations with gas installations and equipment must ensure that this is checked by a CORGI-registered installer at least every year. Any repairs must be carried out promptly, also by a CORGI-registered installer. In some offices maintenance will be the responsibility of the landlord.	Maintenance and repairs of gas equipment are carried out by:
	6.3 BOILERS Arrangements must be in place for regular maintenance and, where necessary, repairs of central heating or hot water boilers by a competent organisation. In some offices maintenance will be the responsibility of the landlord.	Boiler maintenance and repairs are carried out by:
	An adequate number of staff must be able to operate the controls of boilers and related equipment.	

Motor trade

The remaining examples of model assessments are from a manual issued by a national network of garage dealerships. The manual contained the company's health and safety policy and the sections of model risk assessments relating to:

- general topics, e.g. fire, electrical installation, site security;
- service workshops;
- parts departments;
- showrooms and administrative areas;
- bodyshops; and
- petrol storage and sales.

The last two sections only applied to a minority of the company's locations.

The first two examples are both adapted from the *service workshops* section and are parts of the subsections dealing with *hazardous substances* and *lifting equipment*. General guidance is provided on risks and the precautions expected and then the dealership must insert details relating to their own arrangements for inspections, maintenance, statutory examinations and training.

A further example is provided which has been adapted from the section for dealerships with *bodyshops*. This provides details on the risks associated with the use of *radioactive anti-static devices* (which are only used by a limited number of body repair workshops). The assessment identifies the precautions which are necessary in the use of such units and directs users towards complying with some of the detailed requirements of the relevant Regulations.

Details of risks	Expected precautions	Local arrangements
	B 9.4 COMMON EXPOSURES TO HAZARDOUS SUBSTANCES (Use of hazardous substances in body shops is covered in section E of the manual).	
	Vehicle exhaust emissions	
Vehicle exhausts contain a variety of hazardous gases.	Workshops should normally be equipped with exhaust extraction systems or have very good standards of general ventilation. Workshop management is expected to ensure that extraction systems are used by staff where appropriate. The system must be inspected and tested by a competent person at least every 14 months with inspection reports kept on file.	The extraction system is inspected and tested by: Inspection reports are kept by:
	Welding and burning	
All types of burning and welding work produce hazardous gases and fumes.	Only limited amounts of welding and burning work are normally carried out in servicing workshops. Where extraction facilities have not been provided, this work must be carried out in a well-ventilated area and preferably in the open air. PPE must be worn as described in section B11 of the manual. Use of respiratory protection will not normally be necessary.	
	Parts cleaning	
Exposure of the skin to cleaning solvents can lead to dermatitis. Some cleaning solvents also create a significant inhalation risk.	Each branch must have a proprietary self-contained parts cleaning unit which is regularly serviced and maintained. Parts should be cleaned by brush wherever possible. Suitable impermeable gloves must be provided for cleaning work.	The parts cleaning unit is serviced and maintained by:

Further parts of this sub-section deal with:

- petrol and oil;
- battery acid;
- anti-freeze;
- valeting chemicals;
- brake linings (possible asbestos risk);
- air conditioning units (fluorocarbon gases); and
- aerosols.

B12 *Lifting equipment*

Details of risks	Expected precautions	Local arrangements
	B 12.1 TESTING AND EXAMINATION All lifting equipment must have a declaration of conformity or a report of a thorough examination specifying its safe working load, (provided by the suppliers or manufacturers) before it is taken into use. Every item of lifting equipment must carry a clear identifying reference number (or letter).	
The condition of lifting equipment may deteriorate due to damage, misuse, the environment or wear and tear.	Lifting equipment must continue to be thoroughly examined by a competent person (usually an insurance company engineer) throughout its lifespan. The statutory examination periods are normally: 6 months for accessories for lifting such as chains, ropes and lifting tackle (including fabric and wire rope slings, shackles, eye bolts, etc.) 12 months for other lifting equipment (including vehicle hoists, fork lift trucks, chain blocks, electric hoist blocks and engine lifters).	Thorough examinations are carried out by: Examination reports are kept by:
Inadequately trained staff may operate lifting equipment incorrectly or be unaware of good lifting and slinging practice.	*B 12.2 USE OF LIFTING EQUIPMENT* Fork lift trucks may only be operated by persons who have been trained in accordance with the HSE Approved code of practice (ACOP). All staff involved in using lifting equipment must be provided with appropriate in-house training and suitable records of training should always be kept.	The following persons have been trained to ACOP standard in fork lift truck operation:
Equipment may have been damaged since the last time it was used.	Staff should always check the condition of lifting equipment prior to using it. Equipment in poor condition must be taken out of use. (Where lifting equipment is scrapped, its reference number should be recorded so that it is removed from the 'thorough examination' inventory.)	
Overloading of lifting equipment could result in its sudden dramatic failure or in hidden damage.	The weights of the load to be lifted should be established (either by reference to manuals or by a realistic estimate). The safe working load of all items of lifting equipment being used must be adequate for the load.	Lifting equipment training records are kept by:

(Further guidance on good lifting procedure is provided in the remainder of this sub-section.)

E7 *Radioactive anti-static devices*

Details of risks	Expected precautions	Local arrangements
	E 7.1 INTRODUCTION	
Some body shops use radioactive sources in compressed air blow down guns to reduce the problems caused by static electricity. (Suppliers of this equipment will usually give specialist assistance in compliance with the relevant Regulations.)	Any dealership keeping such radioactive sources must be registered with the Environment Agency (EA) and keep a copy of the registration on file.	The Environment Agency registration is kept by:
	The HSE must be notified prior to use of the radioactive source commencing. (In most cases, the suppliers will arrange for EA registration and HSE notification on behalf of the dealership.)	The HSE have been notified by:
	A competent member of staff must be appointed as a radiation protection supervisor (RPS). The RPS must be aware of the contents of this section of the manual and relevant information from the supplier, particularly in respect of methods of use of the unit and the actions necessary in case of its damage or loss.	The radiation protection supervisor is:
	Instructions relating to the use of the unit and actions in case of damage or loss must be prominently displayed in the area where it is used. A record must also be kept of the unit's location. (Some suppliers provide a wall chart for these purposes.)	Instructions about the unit are displayed:

E7 Radioactive anti-static devices

Details of risks	Expected precautions	Local arrangements
	E 7.2 USE OF THE UNIT	
	When the ionising gun is not in use it must be detached from the air line and locked in a storage box. (Any spare cartridge must also be locked in the box.)	The lockable storage box is kept:
	The unit must only be used for its intended purpose. The unit must not be modified, cleaned or disposed of. The foil surface must not be damaged and the nozzle must not be probed with wire or any other pointed instrument.	
	Suppliers' requirements for returning cartridges to them must be complied with.	
	The unit must not be moved to another location without consulting the supplier (further EA registration and HSE notification may be required).	
	The unit should be checked regularly for signs of damage or deterioration.	
	E 7.3 DAMAGE TO OR LOSS OF THE UNIT	
The unit may be damaged by fire, mechanical impact, exposure to corrosive materials, etc. It may also be lost or stolen.	All cases of damage or loss or suspicions of its loss or theft must be reported immediately to the RPS.	
	The RPS is responsible, where necessary, for notifying the suppliers and following their advice. Notification to the EA and/or HSE may also be required. (If the RPS cannot be contacted, advice should be obtained from the suppliers directly.)	The home telephone number of the RPS and the suppliers' emergency number are displayed:
Fire in the area of the unit.	The Fire Brigade must be informed of the unit's presence.	
Fire, mechanical damage or corrosion affecting the unit.	Access to the area must be prevented and advice sought from the suppliers or any other competent source.	
Mechanical or corrosive damage to the foil may release small quantities of radioactive material.	Anyone who has touched the gun, or any object in the immediate area, should wash their hands thoroughly. Further contact should be avoided until competent advice has been obtained.	

7 Specialised risk assessment techniques

In this chapter:

Introduction	**7.1**
Risk rating matrices	**7.2**
Advantages	**7.3**
Disadvantages	**7.4**
More sophisticated approaches	**7.5**
HAZOP	**7.6**
References	**7.7**

Introduction

7.1 The *Management Regulations 1992* resulted in a variety of risk assessment techniques being developed, many of which involved the use of risk rating or quantitative assessment techniques. However, the ACOPs accompanying both the 1992 and the 1999 Regulations have emphasised that the degree of sophistication of the risk assessment should be determined by the level of risk arising from the work activity. The 1999 ACOP provides further detail on this point:

Small businesses presenting few or simple hazards:

- 'risk assessment can be a very straightforward process based on informal judgement and reference to appropriate guidance'.

Intermediate cases:

- risk assessment will need to be more sophisticated;
- specialised advice may be required in some cases; and
- some specialist analytical techniques may be required, e.g. measuring air quality and assessing its impact.

Large and hazardous sites:

- require the most developed and sophisticated risk assessments, particularly where there are complex or novel processes;

- for sites using or storing bulk hazardous substances, large-scale mineral extraction or nuclear plant, risk assessment will be a significant part of the safety case or report and may incorporate such techniques as quantified risk assessment; and
- the '*COMAH 1999*' and other statutory requirements (e.g. in the nuclear industry) may require more specific and detailed assessment techniques.

Risk rating matrices

7.2 Many different risk rating matrices have been developed as an aid to the risk assessment process. Several examples are put forward below whilst the pros and cons of this type of approach are set out in the later sections (see **7.3: ADVANTAGES** and **7.4: DISADVANTAGES**).

Most matrices utilise a simple combination of the likelihood of a hazard having an adverse effect and the severity of the consequences if it did. Some use numbers to produce a risk rating, as in the example below:

Risk assessment matrix			*Likelihood of adverse effect*		
			Unlikely	*Possible*	*Frequent*
			1	2	3
Severity of consequences	Minor	1	1	2	3
	Moderate	2	2	4	6
	Severe	3	3	6	9

The numbers can be replaced by descriptions of the level of risk as shown in the next example:

Risk assessment matrix		*Likelihood of adverse effect*		
		Unlikely	*Possible*	*Frequent*
Severity of consequences	Minor	Low	Low	Medium
	Moderate	Low	Medium	High
	Severe	Medium	High	Very high

A more complex version of this approach was provided in a supplement to the May 1993 issue of *The Safety & Health Practitioner*, the magazine of the Institution of Occupational Safety & Health. This approach requires first the identification of the worst likely outcome in respect of each hazard i.e.:

- a fatality;
- major injury or permanent disability (including permanent ill health);

- minor injury; or
- no injury.

Next a judgement must be made of the probability or likelihood of harm occurring, based on the following table:

Probability/likelihood	*Description*
Likely/frequent	Occurs repeatedly/event only to be expected.
Probable	Not surprised. Will occur several times.
Possible	Could occur sometimes.
Remote	Unlikely, though conceivable.
Improbable	So unlikely that probability is close to zero.

Decisions on what actions (if any) should be taken can then be made by reference to the matrix below:

	Likely	*Probable*	*Possible*	*Remote*	*Improbable*
Fatal	1st	2nd	2nd	3rd	
Major injury/ permanent disability	2nd	2nd	3rd		
Minor injury	3rd	3rd			
No injury					

KEY:

(darkest shading)	1st rank actions
(dark shading)	2nd rank actions
(light shading)	3rd rank actions
(no shading)	Acceptable risk – no action

Alternative versions of this matrix approach further divide the severity of the consequences into two components:

- the severity of the injury – e.g. fatal/serious, moderate, minor; and
- the number of people who could be harmed – e.g. many, more than one, one.

Advantages

7.3 Advocates of risk rating using these types of matrices put forward the following arguments in its favour:

- it ensures that both severity and likelihood are considered;
- some subjectivity is removed from the process; and
- it helps in determining priorities for improvements.

Disadvantages

7.4 Detractors (including the author) feel that the approach is pseudo-scientific and unnecessary:

- its application to every hazard associated with each individual activity or situation can be very time consuming;
- much time can be spent debating risk values, rather than evaluating the effectiveness of controls; and
- some quick and simple measures to improve the control of lesser risks may be overlooked whilst attention is focused on higher-scored risks.

As more organisations gain in experience of actually conducting risk assessments in practice, opinion seems to be swinging against the use of risk-rating matrices and more in favour of the approach described in **CHAPTER 4: CARRYING OUT RISK ASSESSMENTS**. The HSE publication *HSG 183 Five steps to risk assessment: case studies (1998)* does not utilise the matrix approach.

More sophisticated approaches

7.5 The ACOP recognises that more sophisticated approaches to risk assessment must be taken in high risk situations. Many such situations are controlled by the *COMAH Regulations 1999* and HSE guidance on the Regulations (*HSG 190 Preparing safety reports. Control of Major Accident Hazard Regulations 1999 (1999)*) refers to several techniques which may be used, including:

- **HAZOP**
  Hazard and operating studies (HAZOP) were first used by ICI in the 1960s to identify hazards, particularly during the design of chemical plants. The HAZOP approach is described more fully later in the chapter (see **7.6: HAZOP**).
- **FMEA**
  Failure mode and effects analysis (FMEA) is a technique which can be used to calculate the possibility of failure of components in a piece of equipment or a system and thus calculate the possibility of failure of the equipment or system as a whole.

- **Event tree analysis**
  This technique starts from a possible component failure and is intended to identify possible resulting hazards. It is used for rather more complex systems than FMEA.
- **Fault tree analysis**
  Fault trees start from a hazardous outcome and then work downwards to identify the potential causes of such an outcome and the possibilities of these causes occurring. It is a 'top-down' approach as opposed to event tree analysis which involves analysis from the bottom-up.

Detailed description of these techniques is outside the scope of this handbook but is available in other more specialised publications.

The HSE and other organisations have commissioned studies into quantified risk assessment (QRA). In the Foreword to one such paper relating particularly to the nuclear industry (HSE publication *Quantified risk assessment: its input to decision making (1989)*), the HSE state:

> QRA is an element that cannot be ignored in decision-making about risk since it is the only discipline capable, however imperfectly, of enabling a number to be applied and comparisons of a sort to be made, other than of a purely qualitative kind. This said, the numerical element must be viewed with great caution and treated as only one parameter in an essentially judgemental exercise. Moreover, since any judgement upon risk is distributional, risks being caused to some as an outcome of the activity of others, it is therefore essentially political in the widest sense of the word.

Much the same can be said in respect of attempts to quantify risk at far more basic levels.

HAZOP

7.6 As stated previously, HAZOP is a technique that can be used to identify potential hazards and to introduce appropriate precautions to control those hazards. It is particularly useful when designing chemical plants and similar installations, although it can be applied to existing plants. It is best used by a multi-disciplinary team involving persons with expertise in design, commissioning, operations, maintenance and health and safety functions.

The technique identifies hazards through the application of guide words to individual elements of the proposed design. The HAZOP team brainstorms how each guide word might apply to the element under review, identifying deviations from the intended performance, possible causes of these deviations and potential consequences of the deviation. Where a significant risk is identified as a result, actions are specified which should reduce the risk to an acceptable level. These actions may involve modifications to the design, making changes at the installation location, introducing maintenance or inspection routines or adopting specified precautions during operation.

The accompanying table shows how the HAZOP guidewords can be used to identify deviations and is followed by an example of a completed HAZOP worksheet.

HAZOP: Guide words for the generation of process deviations

Guide word	Possible deviations
None	No liquid flow. No electric current. No pressure. Reverse flow. Operational sequence omitted.
More of (or *Less of*)	Quantitative increase (or decrease) in any parameter, e.g. flow, pressure, temperature, electric current, viscosity, volume, weight, dimension.
Part of	Only some of the design intention is achieved, e.g. a change in chemical composition, incomplete reaction.
More than	Something else is present, e.g. impurities in a raw material, gas is present in a liquid (or liquid in a gas).
Other	What else can happen? • instrumentation failure; • sampling activities; • corrosion of components; • reliefs activating, e.g. blow-off valve; • service failure, e.g. cooling water, compressed air; • maintenance activities; or • static electricity generated/discharged.

HAZOP Worksheet

Project	New manufacturing facility – Flixwood				
Section	Solvent pumps and supply pipelines			Sheet 1 of 3	
Brief description of function	Supply of solvent from the tank farm to the manufacturing building			**Date reviewed**	14 Aug 2000
				Date agreed by team	21 Aug 2000
Guideword	**Deviation**	**Possible causes**	**Consequences**	**Action required**	**Responsibility**
More of	Increasedflow.	Higher solvent pressure during start-up.	High turning moment on dog-leg pipe.	Check whether pipe can stand start-up pressure.	Design engineer
			Possibility of rupture.		
None	No flow.	Pump seal leak.	Major leakage in pump area, possibility of fire.	Reliable pump seals already selected.	
				Consider need for remotely operated isolation valve.	Design engineer
				Provide suitable fire fighting equipment in area.	Operations
		Pipeline failure due to impact damage.	Major leakage next to road, high risk of fire.	Protect pipeline with crash barriers.	Design engineer

References

7.7

1	HSG 183	5 steps to risk assessment: case studies, HSE (1998)
2	HSG 190	Preparing safety reports. Control of Major Accident Hazard Regulations 1999, HSE (1999)
3		Quantified risk assessment: Its input to decision making, HSE (1989)

8 Implementation of precautions

In this chapter:

Introduction	**8.1**
Principles of prevention	**8.2**
The management cycle	**8.3**
Planning	**8.4**
Health and safety policy statement	**8.4**
Annual health and safety plans	**8.5**
Development of performance standards	**8.6**
Risk assessments	**8.7**
Organisation	**8.8**
Responsibilities for implementation	**8.8**
Communication and consultation	**8.9**
Advice and information	**8.10**
Control	**8.11**
Procedures and systems	**8.11**
Training programmes	**8.12**
Supervision	**8.13**
Monitoring	**8.14**
Health and safety inspections	**8.14**
Health and safety audits	**8.15**
Other monitoring techniques	**8.16**
Accident and incident investigation	**8.17**
Accident ratio studies	**8.18**
Causes of accidents	**8.19**
Purpose of investigation	**8.20**
Review	**8.21**
Health and safety committees	**8.21**
Management meetings	**8.25**
A continuous cycle	**8.26**
References	**8.27**

Introduction

8.1 The time spent in conducting risk assessments of any type can be regarded as wasted unless the precautions identified as being necessary are actually implemented. Much of this chapter is concerned with the management actions which should be taken to ensure effective implementation – the 'management cycle'. However, first a requirement introduced by the *1999 Management Regulations* will be examined, i.e. that certain principles of prevention must be applied.

Principles of prevention

8.2 *Regulation 4* of the *Management Regulations 1999* states:

> Where an employer implements any preventive and protective measures he shall do so on the basis of the principles specified in Schedule 1 to these Regulations.

Schedule 1 specifies the general principles of prevention set out in *Article 6(2)* of *European Council Directive 89/391/EEC*. These are:

(a) avoiding risks;

(b) evaluating the risks which cannot be avoided;

(c) combating the risks at source;

(d) adapting the work to the individual, especially as regards the design of workplaces, the choice of work equipment and the choice of working and production methods, with a view, in particular, to alleviating monotonous work and work at a predetermined work-rate and to reducing their effect on health;

(e) adapting to technical progress;

(f) replacing the dangerous by the non-dangerous or the less dangerous;

(g) developing a coherent overall prevention policy which covers technology, organisation of work, working conditions, social relationships and the influence of factors relating to the working environment;

(h) giving collective protective measures priority over individual protective measures; and

(i) giving appropriate instructions to employees.

These principles – particularly avoiding risks, combating risks at source and giving collective protection priority over individual protective measures – have been accepted for many years in health and safety. In some UK Regulations a hierarchical approach is required. For instance, in the *COSHH Regulations 2002* (see **CHAPTER 9: COSHH ASSESSMENTS**) exposure of employees to hazardous substances must be prevented where reasonably practicable. If this is not the case, exposure must be adequately controlled – according to a clear order of preference:

1 prevention of exposure (by elimination or substitution);

2 adequate control by work processes, systems and engineering controls (e.g. enclosure) and use of suitable work equipment and materials;

3 adequate control of exposure at source (by ventilation or organisational measures); and

4 adequate control, through PPE.

A similar approach is taken in *Regulation 11* of the (PUWER) *1998* (as amended by the *Health and Safety (Miscellaneous Amendments) Regulations 2002*). This requires effective measures to be taken to prevent access to any dangerous part of machinery (or any rotating stock bar) or to stop its movement before any part of a person enters a danger zone. In this case, the following measures must be adopted 'where and to the extent that it is practicable to do so' in this order of preference:

1 fixed enclosing guards;

2 other guards or protection devices, e.g. interlocked guards, photoelectric beams;

3 protection appliances, e.g. jigs, holders or push-sticks; and

together with 'the provision of such information, instruction, training and supervision, as is necessary.'

Note that this hierarchy must be more strictly applied than in the *COSHH Regulations* since the qualifying term used is 'practicable' as opposed to 'reasonably practicable'.

Whilst the *HSE Guidance to Regulation 4* of the *Management Regulations 1999* clearly indicates a preference for a similar hierarchical approach to be taken, it also accepts that this will not always be possible. In *Paragraph (31)* the guidance states: 'These are general principles rather than prescriptive requirements. They should, however, be applied wherever it is reasonable to do so. Experience suggests that, in the majority of cases, adopting good practice will be enough to ensure risks are reduced sufficiently.'

The management cycle

8.3 *Regulation 5* of the *Management Regulations 1999* requires employers to take effective steps to implement the precautions identified by the risk assessment as being necessary – the theory must be translated into practice in the workplace. *Paragraph (1)* of *Regulation 5* states:

> Every employer shall make and give effect to such arrangements as are appropriate, having regard to the nature of his activities and the size of his undertaking for the effective planning, organising, control, monitoring and review of the preventive and protective measures.

Paragraph (2) of *Regulation 5* requires employers with five or more employees to record these arrangements.

Such a 'management cycle' has long been applied to other areas of business activity, such as finance, but relatively few employers have utilised it in relation to health and safety. Managers have often stated their good intentions but have not always set up the organisational structure and control to implement those intentions and have failed to monitor what is actually happening in the workplace.

The cycle can be applied to an employer's overall approach to health and safety:

- Plan – through the statement of intent within the health and safety policy;
- Organise – by allocating responsibilities for implementing the policy and making the necessary resources available;
- Control – through application of relevant management systems and techniques and the use of performance standards;
- Monitor – through health and safety audits and inspections; and
- Review – in health and safety committee and management meetings.

The cycle can also be applied in other ways to ensure that systems or procedures are implemented effectively (this is illustrated in relation to health and safety inspections later in the chapter). It can also be applied to specific types of precautions such as the provision of guarding for dangerous parts of machinery.

- Plan by stated objectives, e.g. compliance with legislation (e.g. PUWER 1998) or related European or British standards;
- Organise – through having staff competent to interpret and apply such standards;
- Control – by procedures for selecting new equipment or modifying existing equipment and arrangements for supervisory staff to ensure that guards are used as appropriate;
- Monitor – through the health and safety inspection programme; and
- Review – via problem solving or other meetings which result in actions such as redesign of guards or the re-training of supervisors or employees.

In the remainder of this chapter each of the five stages of the management cycle will be examined in more detail. It should perhaps be emphasised that the various elements which make up health and safety management programmes can straddle more than one stage of the management cycle. Whilst a health and safety committee will *review* the effectiveness of health and safety measures, its members should also be expected to *monitor* precautions – both through formal inspection programmes and by informal observations during their daily work. The ACOP to the *Management Regulations* states that whilst more complex health and safety management systems may be appropriate for large complicated organisations, the principles of good health and safety management are the same, irrespective of an organisation's size. It refers to the key elements of effective systems as contained in *Successful Health and Safety Management* (**REF. 1**) and *BS 8800* (**REF. 2**).

Planning

Health and safety policy statement

8.4 *Section 2(3)* of *HSWA 1974* requires all employers with five or more employees to have a written statement of their health and safety policy. This should have three components – a statement of intent, responsibilities within the organisation for implementing the policy and the arrangements for implementing the policy. All of these form part of the planning process:

- Statement of intent – where does the organisation plan to be on health and safety?
- Responsibilities – who should be made responsible for implementing actions within the organisation? (equipping people to carry out those responsibilities is considered later under 'organisation'); and
- Arrangements – what procedures and mechanisms need to be established to deliver the desired standards? (the nature of such procedures is considered later under 'control').

Annual health and safety plans

8.5 Whilst many organisations achieve good health and safety standards, none are perfect – the minority that think they are perfect are deluding themselves! All employers should seek constant improvement in their management systems and should also recognise that they must take account of changes taking place both externally (new Regulations, changes in ACOPs and HSE guidance) and internally (new equipment, processes, materials and organisational restructuring). Many employers now choose to prepare an annual health and safety plan to achieve desired improvements or cope with changes. Such plans should of course be accompanied by identification of responsibilities, timescales and resources for their implementation.

Development of performance standards

8.6 Objectives contained in policy statements or annual health and safety plans relating to improved accident performance are just wishful thinking unless the actions intended to achieve the improvement are properly identified. These actions must be:

- Measurable – a defined activity to be carried out by a specific date or to a specified frequency; and

- Realistic – capable of being achieved by those responsible within the timescale specified, taking into account the availability of resources and other prevailing circumstances.

Such standards can be built into annual plans

e.g. – Refresher training in manual handling (to a specified syllabus) will be delivered to all warehouse staff by the safety officer during the second quarter.

They can also be incorporated into procedures or systems

e.g. – All new employees and transferees will receive health and safety induction training (of a specified content) from their manager or supervisor during their first day in their new department.

e.g. – A health and safety inspection will be carried out in each department every month by a supervisor and a safety representative.

Risk assessments

8.7 The importance of planning the risk assessment programme was stressed in **CHAPTER 4: CARRYING OUT RISK ASSESSMENTS**. Once the initial programme has been completed, planning of how assessments are to be reviewed and revised must also take place.

Organisation

Responsibilities for implementation

8.8 Where individuals are allocated responsibility for implementing elements of the health and safety programme they must be capable of accepting such responsibilities. This applies whether the responsibilities result from the health and safety policy, an annual health and safety plan or a need to carry out risk assessments. For example:

- A newly appointed supervisor is likely to need training on relevant health and safety legislation and his responsibilities under it, together with procedures such as permits to work, accident investigation or emergency evacuation that he must now implement;

- A safety officer may need training in instructional techniques, even before delivering training on a subject he is already familiar with, such as manual handling; and
- A risk assessment team is likely to require training both on the legal background to their work and on assessment techniques before embarking on their programme.

Communication and consultation

8.9 The value of health and safety committees in respect of monitoring and review will be stressed later. However, such a committee or an alternative means of consulting employees must first be established. Employers have legal duties, to consult employees or their representatives, imposed on them by the *Safety Representatives and Safety Committees Regulations 1996* (**REF. 3**) and the *Health and Safety (Consultation with Employees) Regulations 1996* (**REF. 4**). Essential health and safety information must also be communicated to employees through either induction programmes, safety handbooks, newsletters, notice boards briefing sessions. All of these activities require organisational effort on the part of the employer.

Advice and information

8.10 Employers have a duty imposed on them through *Regulation* 7 of the *Management Regulations 1999* to have competent health and safety assistance (see **CHAPTER 2: WHAT THE MANAGEMENT REGULATIONS REQUIRE**). The employer may decide to carry out this role himself, appoint one or more employees to provide the assistance or engage the services of an external consultancy. If the latter course is adopted, the employer will still need to organise the selection of a suitable consultancy (**REF. 5**). Alternatively the employer or the appointed employee(s) will require an appropriate level of training in order to fulfil the role. There will also be a need for ongoing information to ensure the organisation keeps abreast of developments in health and safety – this might be provided through specialist magazines, information update subscription services or the Internet.

Control

Procedures and systems

8.11 Control is often established through the creation of formal procedures and systems for delivering key elements of the health and safety programme. Such procedures will provide detail of the 'arrangements' parts of the health and safety policy. The need for procedures will depend upon the size and complexity of the organisation, and topics might include:

- Health and Safety Inspections;
- Accident and Incident Investigation;

- Fire Evacuation;
- Other Emergencies e.g. bomb threats, hazardous substance leaks;
- Medical Screening;
- Permits to Work;
- Selection, Purchase and Issue of PPE;
- Purchase of New Equipment;
- Engineering Projects;
- Selection and Management of Contractors;
- First Aid.

These would be in addition to any operational or production procedures which might include significant health and safety content (as in the example provided at the end of **CHAPTER 5: ASSESSMENT RECORDS**).

Training programmes

8.12 Control can also be established through specifying the content of training programmes throughout the organisation. Such training may relate to:

- Induction – a defined programme for new employees (inc. temps) and transferees;
- Processes – utilising the type of procedure at the end of **CHAPTER 5: ASSESSMENT RECORDS**;
- Procedures – e.g. accident investigation, permit to work issue/receipt;
- Equipment – e.g. driving fork lift trucks, operating cranes;
- Activities – e.g. manual handling, entering confined spaces, working at heights;
- Status – e.g. programmes for newly appointed supervisors;
- Changes in legislation – updating of specialists or senior managers;
- Health and safety awareness – for managers, supervisors or employees generally.

In many cases the detailed health and safety content of training programmes can be derived from a process of risk assessment. This is particularly the case in respect of induction training. It is a relatively simple process to extract the key pieces of information a new employee needs by using the risk assessment for the department or section in question. Some training will need to be accompanied by testing in order to confirm that the employee has achieved a desired level of understanding or skill.

Supervision

8.13 Good standards of supervision are also essential in achieving control of what happens in the workplace. Supervisors have an important role to play in:

- communicating and consulting with employees;
- delivering induction and operator training;
- implementing procedures;
- enforcing standards, e.g. PPE requirements, speed limits;
- identifying deficiencies in procedures (or the need for new ones).

This latter task is part of the important assistance that supervisors can provide in the monitoring of health and safety standards.

Monitoring

Health and safety inspections

8.14 Conducting regular health and safety inspections is an important means of ensuring that standards are monitored systematically. It is important that inspections take account of the way people are working rather than just paying attention to equipment and premises. It is well established that 'unsafe acts' (e.g. non–compliance with PPE standards, speeding, removing guards, cutting corners on procedures) are a much more common cause of accidents than 'unsafe conditions'. Indeed many unsafe conditions (e.g. missing guards, damaged equipment, blocked fire exits) result from earlier unsafe acts.

Where trade unions are formally recognised by employers, union-appointed safety representatives have a right to carry out inspections at least every three months of areas where their members work (**REF. 3**). Many employers arrange for supervisors or managers to carry out inspections jointly with these safety representatives and many more consider it good practice to involve employees in inspections, even when there is no formal recognition of a union.

Whilst those carrying out inspections should be alert for any possible problems, there will inevitably be some issues which are of more importance in specific workplaces. Inspection checklists can be derived from risk assessment records which highlight these important issues. These might relate to:

- compliance with PPE requirements;
- maintenance of fire exit routes or self-closing fire doors;
- presence of or adjustment of guards;
- manual handling practices;

- housekeeping standards or condition of floors;
- driving standards.

An example of an inspection checklist relating to an office environment is provided later in the chapter. This is accompanied by a sheet which can be used to track action points resulting from the inspection.

The management cycle can be applied to the implementation of a health and safety inspection programme:

- **Plan**
  - a statement of intent to carry out regular inspections and involve employee representatives;
- **Organise**
  - provide training in inspection technique for those who are involved;
  - identify areas to be inspected, e.g. by splitting a large workplace into inspection units;
  - define the frequencies for inspections to take place (these may vary between areas with different levels of risk);
  - identify who is responsible for making the inspections, e.g. the supervisor and safety representative jointly;
  - provide area specific inspection checklists;
  - establish a mechanism for allocating responsibility for remedial actions.
- **Control**
  - develop a procedure defining all the above arrangements;
  - provide reminders for inspection participants;
  - ensure participants have time available to carry out inspections.
- **Monitor**
  - check that inspections are being made as scheduled;
  - review the quality of inspection reports;
  - check whether remedial actions are being implemented.
- **Review**
  - reasons for any shortcomings in the system, e.g. failure to carry out inspections due to sickness absence or pressure of other duties, ineffective communication of remedial actions required;
  - identify means of overcoming these differences.

MIDSHIRES DEVELOPMENT AGENCY OFFICE HEALTH AND SAFETY INSPECTION Office Location: Inspection by: Date:		
Risk Topic	**Specific items**	**Comments**
1. FIRE	Evacuation routes available SC doors not wedged open Alarm tests taking place Extinguishers in position	
2. ACCIDENTS, etc.	First aid equipment	
3. VISITORS AND CONTRACTORS	Signing in/out arrangements Control of contractors	
4. ACCESS (INTERNAL AND EXTERNAL)	Corridors, walkways clear Access to shelves, cupboards, etc. External footpaths, roadways	
5. OFFICE & KITCHEN EQUIPMENT	Condition of equipment Cables, plugs, sockets, etc.	
6. SERVICES	Fixed electrical installation Access to isolators, switches, etc.	
7. HAZARDOUS SUBSTANCES	Control of cleaning materials Gloves available/used	
8. MANUAL HANDLING	Archives, stationery storage Access to shelves	
9. DISPLAY SCREEN EQUIPMENT	DSE Workstations	
10. OTHER ITEMS		

DETAIL ACTION POINTS OVERLEAF

HEALTH AND SAFETY INSPECTION ACTION POINTS

Date of inspection:

* **Priority Code: A** – Urgent; **B** – Important; **C** – Routine

Ref	Action Point	Priority Code*	Action By	Details of Progress	Initials

Health and safety audits

8.15 The terms 'audits' and 'inspections' are often used interchangeably but in health and safety terminology they have gradually developed to have distinct meanings.

- *Health and safety inspection* – a check of physical working conditions, equipment, working practices and behaviour of employees (and others), usually carried out over a relatively short period of time.
- *Health and safety audit* – a much more detailed and comprehensive evaluation of management systems. This is likely to involve checking on policies and procedures, examining records, interviewing employees and carrying out some sample inspections – to determine whether the systems are working in practice.

Several commercial health and safety auditing systems are available including the International Safety Rating System (ISRS), the British Safety Council's 5 Star Audit and HASTAM's CHASE. However, these may have limitations for some organisations either because they are products of another country's health and safety culture or they do not deal with all of the aspects of health and safety management appropriate to the organisation in question. As a result many organisations have developed their own bespoke auditing systems.

Another important consideration is whether the audit is to be purely qualitative or quantitative as well. Quantification can aid in comparing performances between different locations or in measuring progress (or deterioration). However, there is a danger of the emphasis of the audit switching to points chasing or quibbling about the auditor's judgement rather than focussing on the systematic evaluation of safety management systems.

The best auditing systems are those which place only limited reliance on the individual auditor's judgement. The questions within the audit should preferably be capable of being answered with a simple yes or no and should by their nature be indicative of what is required in a correctly functioning safety management system. Such systems, if properly applied, can be used just as productively for self-audits as by external auditors. In either case the auditor should require confirmatory evidence that a system is in place and working, rather than accepting unsubstantiated assurances.

Topics which may be appropriate for health and safety auditing are:

- Risk Assessments;
- Operating Procedures;
- Emergency Procedures;
- Training;
- Occupational Health;
- Inspections and Audits (in-house);

- Accident and Incident Investigation;
- Control of Contractors;
- Communication with Employees;
- Engineering and Purchasing Controls.

Health and safety-related procedures can also be included within audits carried out of procedures generally as part of ISO 9000 control systems.

Other monitoring techniques

8.16 Informal monitoring of health and safety standards may be carried out by managers, health and safety specialists or safety representatives. Other more specialised techniques include:

- *Safety surveys* – a detailed examination of a specific part of a management system (e.g. quality of accident investigation reports) or a specific activity or item of equipment (e.g. use of ladders).
- *Safety sampling* – a detailed inspection of a small work area.
- *Safety tours* – a walk through evaluation of a workplace to form a general impression of standards (often conducted by a senior manager).
- *Safety observations* – a detailed observation of the work methods and equipment used in carrying out a specific task.
- *Compliance surveys* – a check on a readily observable aspect of employee behaviour, e.g. compliance with speed limits or a PPE requirement.

Accident and incident investigation

8.17 Whilst inspections and audits provide pro-active means of monitoring health and safety performance, the investigation of accidents and incidents constitutes an important reactive monitoring technique.

Accident ratio studies

8.18 Heinrich and others have studied the relationship between major injury, minor injury and non-injury accidents. Heinrich's data showed that for every major or lost time injury there were 29 minor injuries and 300 non-injury incidents. Other researchers (Bird in the USA and Tye/Pearson in the UK) have demonstrated similar ratios (**REF. 1**). As an illustration, if a man slips on a patch of spilled oil, he may be unhurt, he may damage clothing or equipment, he may break his arm or he may fracture his skull and die. As the studies show, most of the time the man is lucky and the consequences of his accident are small. However, the potential for injury should always be considered – effective accident prevention and loss control should concentrate on the causes of the accident. The reasons for the spilled oil should be sought and remedied in order to remove the potential for a serious accident.

Causes of accidents

8.19 (For brevity, the term 'accidents' is also intended to embrace non-injury incidents.)

Most accidents have multiple causes, with several circumstances combining to produce an unwanted result. Frank Bird in his book *Practical Loss Control Leadership* also developed Heinrich's domino model of loss causation, which consists of five toppling dominoes:

- The final domino to fall represents a *loss* (e.g. personal injury, equipment damage, etc.);
- This is due to a specific *incident* (e.g. a slip on a patch of oil), represented by the fourth domino;
- The incident has *immediate causes* (e.g. a leaky vehicle or a failure to clear up the spill), which is the third domino;
- Behind the immediate causes is the second domino – *basic causes* (e.g. inadequate maintenance, lack of oil absorbent materials, absence of clean up procedures);
- These basic causes are symptomatic of a *lack of control* – the first domino. A well-managed workplace has well-regulated maintenance procedures, an adequate supply of materials, etc. If the business is correctly managed the first domino will not topple onto the others and there will be no resultant loss.

Purpose of investigation

8.20 The purpose of accident and incident investigation systems should always be to identify shortcomings in the management of health and safety. Organisations should learn from the large numbers of non-injury incidents and minor accidents which the accident ratio studies demonstrate are occurring, and utilise them to prevent future accidents, possibly with more serious consequences in terms of injury or damage.

Whilst there is also a need to investigate in order to submit reports under RIDDOR or in order to provide information in relation to future compensation claims, investigation procedures which are driven purely by RIDDOR and insurance company needs will not be successful in reducing accidents. The objective should always be never to waste an accident.

Investigation procedures

Larger organisations should have a formal procedure for investigating accidents and incidents while smaller employers should at least have clearly established local arrangements.

These should include the following elements:

- What should be investigated
  - Whether a particular set of circumstances causes an 'Accident', damage to plant or materials, a 'Dangerous Occurrence' or just a near miss, may be a matter of chance;
  - Significant injuries and damage incidents must always be thoroughly investigated, but so also should many minor accidents or near misses, especially those with potentially more serious consequences.
- The purpose of investigation

  Investigation should be concerned with:
  - Finding out *what* happened – notification under RIDDOR may be necessary and accurate facts need to be recorded in case of possible compensation claims;
  - Finding out *why* it happened – immediate causes, e.g. faulty equipment, unsafe methods, and basic causes, e.g. inadequate maintenance, lack of training or supervision,
  - *Preventing* it happening again – getting things fixed, improving procedures, changing employees' behaviour or attitudes – or arranging for others to do so;
  - Procedures should emphasise each of these aspects.
- Investigation timescale
  - Investigations should normally start as soon as possible after the incident and a minimum timescale for submission of at least an interim report should be set.
- Responsibility for investigation
  - This will normally rest with front-line supervision although senior managers or specialists may well become involved in more serious or complex incidents.
- Reports and remedial action
  - It must be clearly established who is to receive and review reports and who is to allocate responsibility for and follow up the necessary remedial action – the most important step in the procedure from the accident prevention viewpoint.

Investigation technique

Good investigation technique should involve the following elements:

- Observing
  - Looking at the scene and the surrounding area (not being an armchair investigator).

- Interviewing
  - the injured person and/or witnesses (preferably separately);
  - at the scene if possible;
  - noting down beforehand some key questions to be answered;
  - asking open-ended questions in a friendly manner;
  - keeping an open mind.
- Involving others
  - Seeking advice from specialists (engineers, chemists) or others who may be able to contribute.
- Drawing Conclusions
  - On what happened, why it happened and how it might be prevented from happening again.
- Reporting
  - See below.
- Taking action
  - The Investigator should take whatever action he or she is able to and press others to act on recommendations made.

Investigation reports

- Report forms should be designed to contain all the relevant details of the accident or incident
  - who or what was involved;
  - details of any injury or damage;
  - the date and time;
  - the name and signature of the person investigating.
- Sufficient space should be allocated to record
  - *what* happened – a descriptive report;
  - the investigator's conclusions as to *why* it happened, including both immediate and basic causes;
  - recommendations to *prevent* it happening again.

A sample investigation report form is provided later within this chapter. (It should be noted that some organisations require reports to be submitted to Departmental Managers before or in addition to health and safety specialists.)

ACCIDENT & INCIDENT INVESTIGATION REPORT

Department ______________________		Section ______________________	
Person involved	Surname ____________	First name(s) ______________________	
Status of person	Employee ☐	Visitor ☐	Member of Public ☐
(tick as appropriate)	Temp. ☐	Contractor ☐	Other (state)______ ☐
Date	Time	Location	
Details of any injuries (inc. first aid treatment) or damage			
Description of incident – What Happened A sketch or photo may be helpful. Continue on a separate sheet if necessary.			
Contributory causes – Why you think it happened			
What action has been taken (or will be taken) to prevent similar incidents			
Person submitting report	Name	Position	Date
FOR COMPLETION BY THE HEALTH AND SAFETY OFFICER Further action required / Progress of remedial action / Other comments Action was completed on		Name	

Review

Health and safety committees

8.21 In large- and medium-sized organisations the establishment of an effective joint health and safety committee should constitute an important part of the employer's health and safety programme, not just in reviewing the management of health and safety but in providing a forum for:

- consultation on the existence of risks and the effectiveness of precautions;
- sharing knowledge and experience;
- providing different perspectives on health and safety issues;
- giving greater awareness of what is happening within the workplace;
- achieving wider ownership and commitment.

Composition of the committee

8.22 The composition of the Committee should be a matter for local agreement, whether or not a Committee has been formally requested by union Safety Representatives (**ref. 3**). Care should be taken that the Committee has a good balance of representation without becoming too large.

Membership should include:

- Managers and supervisors;
- Health and safety specialists;
- Other specialists, e.g. engineers, chemists;
- Employee representatives.

The employee representatives will often be union Safety Representatives (or other elected representatives), but this need not automatically be the case – other individuals may also be able to contribute.

Committee activities

8.23 An effective Committee can assist the employer at all the stages of the safety management cycle (Plan, Organise, Control, Monitor, Review). Committees are normally active in the following areas:

- reviewing accident and incident reports;
- overseeing the health and safety inspection programme;
- monitoring remedial actions resulting from both of the above;

- reviewing health and safety audits;
- reviewing accident statistics;
- reviewing occupational hygiene survey results;
- planning health and safety initiatives;
- reviewing new and existing procedures and arrangements;
- monitoring the effectiveness of health and safety training;
- reviewing communications from, and actions of, inspectors (HSE/local authority);
- providing feedback on the suitability of PPE;
- assisting in communications and publicity.

Conduct of meetings

8.24 The principles behind holding effective Safety Committee meetings are little different from those governing other meetings:

- there must be a designated Chairperson
  - often a senior manager, although this is not essential.
- someone must be responsible for taking minutes;
- meetings must be held at an agreed frequency
  - often monthly, although 2 or 3 per month may be appropriate for smaller organisations.
- meeting dates must be planned well in advance
  - cancellations or postponements should be avoided.
- a detailed agenda should be circulated to members well in advance;
- votes should be avoided (the Committee should primarily be a consultation and review body, ultimate responsibility rests with the employer);
- minutes should be circulated to members soon after meetings;
- responsibility for implementing actions should be clearly identified;
- Committee activities should be widely publicised
  - e.g. minutes on notice boards and emailed to staff, extracts in newsletters.

Management meetings

8.25 The review of standards of health and safety management may also take place within separate management meetings. Some members of management and health and safety specialists may feel less constrained in such an environment. They may be rather more frank in their appraisal than in a joint forum. Many of the topics appropriate for joint committees will also be appropriate for management meetings. Developing annual health and safety action plans which have the full commitment of the management team and reviewing their subsequent progress are important management activities.

A continuous cycle

8.26 The management cycle should be regarded as continuous, not linear. The review of deficiencies, whether in a joint Health and Safety Committee or in a management only meeting should eventually result in plans for rectifying the deficiencies and the whole cycle should start again.

References

8.27

1	HSG 65	*Successful Health and Safety Management* (HSE 1997)
2	BS 8800	Guide to occupational health and safety management systems British Standards Institute
3	L87	Safety Representatives and Safety Committees Regulations (HSE 1996)
4	L95	A guide to the *Health and Safety (Consultation with Employees) Regulations 1996* (HSE 1996)
5	INDG 322	Need help on health and safety? – free HSE leaflet 2000

9 COSHH Assessments

In this chapter:

Introduction	**9.1**
How substances cause harm	**9.2**
Entry routes	**9.3**
Harmful effects	**9.4**
The COSHH Regulations summarised	**9.5**
Planning and preparing for the assessment	**9.6**
What the Regulations require	**9.6**
Who will carry out the assessments?	**9.7**
How will the assessments be organised?	**9.8**
Gathering information	**9.9**
Prevention or control of exposure	**9.10**
Hierarchy of measures	**9.11**
Prevention of exposure	**9.12**
Control measures, other than PPE	**9.13**
Control using PPE	**9.14**
Adequate control	**9.15**
Carcinogens, mutagens and biological agents	**9.16**
Making the assessment	**9.17**
Observations	**9.18**
Discussions	**9.19**
Sources of help	**9.20**
Further tests and investigations	**9.21**
Preparation of assessment records	**9.22**
Sample assessment records	**9.23**
After the assessment	**9.24**
Review and implementation of recommendations	**9.24**
Use of control measures	**9.25**
Maintenance, examination and testing of control measures	**9.26**
Monitoring exposure at the workplace	**9.30**
Health surveillance	**9.35**
Information, instruction and training	**9.36**
Accidents, incidents and emergencies	**9.37**
Review of assessments	**9.38**

Some pitfalls	**9.42**
The data sheet library	**9.43**
Armchair assesments	**9.44**
Overskill	**9.45**
Not saving the wood for the trees	**9.46**
Slaves to record systems	**9.47**
References	**9.48**

Introduction

9.1 The *Control of Substances Hazardous to Health Regulations* (the *COSHH Regulations*) were first introduced in 1988 but have been amended and changed many times since. The present *COSHH Regulations* came into operation in 2002 but have been subject to subsequent amendments, with the introduction of workplace exposure limits (*Regulation 7*) being the most notable change.

They are the principal Regulations affecting the use of hazardous substances, although there are separate Regulations dealing with asbestos and lead. The Regulations follow general principles which can be applied to most occupational health risks.

- *Assess the risk* – What substances are in use, in what ways and to what extent.
- *Eliminate the risk* – If possible, stop using the hazardous substance or replace it by a less hazardous one, e.g. substitute solvent-based paints by water-based ones.
- *Provide controls* – Control at source, e.g. by enclosure or local exhaust ventilation (LEV) is preferable, although it may be necessary to rely on PPE.
- *Maintain the controls* – Control measures, e.g. LEV and PPE must be checked regularly and maintained in good condition.
- *Monitor* – Check on working conditions and working practices regularly. Atmospheric sampling or health surveillance may be necessary.
- *Inform employees* – About the risks from substances they are using, the precautions they should be following, how to use LEV and PPE etc.

Full details of the 2002 Regulations together with an ACOP and guidance are available in an HSE booklet (**REF. 1**). New editions are published periodically to take account of subsequent amendments.

How substances cause harm

9.2 Before carrying out assessments under the *COSHH Regulations* it is necessary to have an understanding of how hazardous substances can harm the body.

Entry routes

9.3 Hazardous substances may be present in different forms such as:

- Solids
- Dusts
- Smoke
- Fumes
- Liquid
- Mists and aerosols
- Vapour
- Gas.

These can cause harm to the body by three main entry routes:

1. *Inhalation* – The hazardous substance may damage the lungs or some other part of the respiratory system. Inhalation also allows substances to enter the bloodstream and thus affect other parts of the body.
2. *Ingestion* – Accidental ingestion is always possible, especially if containers are not correctly labelled. Eating, drinking and smoking in the workplace introduce risks of small quantities of hazardous substances being inadvertently ingested.
3. *The skin* – Substances can damage the skin (or eyes) through their corrosive or irritant effects. Some solvents can enter the body by absorption through the skin. Entry through cracks or cuts in the skin must also be considered.

Harmful effects

9.4 The detailed consideration of how hazardous substances can harm the body is a matter for more specialist text books. However, the list below illustrates the wide range of problems which may be caused.

- Respiratory problems
  - Pneumoconiosis – e.g. asbestosis, silicosis, byssinosis, siderosis;
  - Respiratory irritation – caused by inhaling acid or alkali gases or mists;
  - Asthma – sensitisation of the respiratory system can be caused by many substances, e.g. isocyanates, flour, grain, hay, animal fur, wood dusts;
  - Respiratory cancers – e.g. from certain chemicals, asbestos, pitch and tar;
  - Metal fume fever – flu-like condition caused by zinc fumes.

- Poisoning – acute (short-term) or chronic (long-term) poisoning may be caused by
  - Metals and their compounds – e.g. lead, manganese, mercury, beryllium, cadmium;
  - Organic chemicals – affecting the nervous system, the liver, the kidneys or the gastro-intestinal system;
  - Inorganic chemicals – as above plus asphyxiant effects, e.g. from carbon monoxide.
- Skin conditions
  - Dermatitis
    - caused by primary irritants, e.g. damage to skin tissue by acids or alkalis or removal of natural oils by solvents, detergents, etc.;
    - due to sensitising agents, e.g. isocyanates, solvents, foodstuffs
  - *Skin cancer* – caused by pitch, tar, soot, mineral oils etc.
- Biological problems – from contact with animals, birds, fish (including their carcasses and products) or with micro-organisms from other sources
  - Livestock Diseases – e.g. anthrax or brucellosis;
  - Allergic alveolitis – from mould or fungal spores present in grain, etc.;
  - Viral hepatitis – usually from contact with blood or blood products;
  - Legionnaire's disease – caused by inhaling airborne water droplets containing the legionella bacteria;
  - Leptospirosis – Weil's disease, from contact with urine from small mammals, e.g. rats.

Information on the harmful effects associated with individual products should be available on the packaging or in a data sheet provided by the manufacturer or supplier.

The COSHH Regulations summarised

9.5 The definition of *substance hazardous to health* is contained in *Regulation 2(1)* of the Regulations and includes:

- substances designated as very toxic, toxic, corrosive, harmful or irritant under product labelling legislation (the *CHIP Regulations*) – these substances should be labelled with the standard orange and black symbols;
- substances for which the Health and Safety Commission has approved a workplace exposure limit;
- biological agents (micro-organisms, cell cultures or human endoparasites);

- other dust of any kind, when present at a concentration in air equal or greater to 10 mg/m^3 (inhalable dust) or 4 mg/m^3 (respirable dust) as a time-weighted average over an 8-hour period;
- any other substance creating a risk to health because of its chemical or toxicological properties and the way it is used or is present at the workplace.

Given this widely drawn definition, in any cases of doubt it is advisable to treat the *COSHH Regulations* as applying.

The Regulations place duties on employers in relation to their employees and also on employees themselves (*Regulation 8*). The self-employed have duties as if they were both employer and employee. *Regulation 3(1)* extends the employer's duties 'so far as is reasonably practicable' to 'any other person, whether at work or not, who may be affected by the work carried out'. Consideration must be given to:

- contractors, visitors or joint occupants;
- neighbours and passers-by;
- members of the public – in public places and buildings;
- customers and service users.

Some of these (e.g. children) cannot be expected to behave in the same way as employees or contractors. An early prosecution under the *COSHH Regulations* was of a doctor's surgery when a young child gained access to an insecure cleaner's cupboard and drank the contents of a bottle of carbolic acid, suffering serious ill effects.

The main requirements of the Regulations are summarised below. Most will be explained in more detail later. Full details of the regulations and associated ACOPs are provided in a single HSE booklet (**REF. 1**).

- *Prohibitions relating to certain substances (Regulation 4)* – the manufacture, use, importation and supply of certain substances is prohibited (further details are contained in *Schedule 2* to the Regulations).
- *Application of Regulations 6–13 (Regulation 5)* – exceptions are made from the application of the Regulations where other more specific Regulations are in place. These relate to respirable dust in coal mines, lead, asbestos, radioactive, explosive or flammable properties of substances, substances at high or low temperatures or high pressure and substances administered in the course of medical treatment.
- *Assessment (Regulation 6)* – an assessment of health risks must be made to identify steps necessary to comply with the Regulations and these steps must be implemented.
- *Prevention or control of exposure (Regulation 7)* – this must be achieved through elimination or substitution or the provision of adequate controls, e.g.

enclosure, LEV, ventilation, systems of work, PPE. This Regulation requires a 'hierarchical' approach to be taken – this is explained later in the chapter.

- *Use of control measures, etc. (Regulation 8)* – employees must make 'full and proper use of control measures, PPE, etc.', and employers 'take all reasonable steps' to ensure they do.
- *Maintenance, examination and testing (Regulation 9)* – control measures, including PPE, must be maintained in an efficient state. Controls must also be examined and tested periodically.
- *Monitoring exposure at the workplace (Regulation 10)* – workplace monitoring may be necessary to ensure adequate control or protect health.
- *Health surveillance (Regulation 11)* – this may be required in some circumstances.
- *Information, instruction and training (Regulation 12)* – must be provided for persons who may be exposed to hazardous substances.
- *Arrangements to deal with accidents, incidents and emergencies (Regulation 13)* – this new Regulation, introduced in 2002, requires emergency procedures to be established and information on emergency arrangements to be provided.

Most of these Regulations are explained in greater detail later in the chapter.

Planning and preparing for the assessment

What the Regulations require

9.6 The wording of *Regulation 6* of the *COSHH Regulations* changed considerably in the 2002 version of the Regulations, with *Sub-paragraph (2)* making it much more specific on the factors which must be taken into account during the assessment. The Regulation now states:

(1) An employer shall not carry on any work which is liable to expose any employees to any substance hazardous to health unless he has–

(a) made a suitable and sufficient assessment of the risk created by that work to the health of those employees and of the steps that need to be taken to meet the requirements of these Regulations; and

(b) implemented the steps referred to in sub-paragraph (a).

(2) The risk assessment shall include consideration of–

(a) the hazardous properties of the substance;

(b) information on health effects provided by the supplier, including information contained in any relevant safety data sheet;

(c) the level, type and duration of exposure;

(d) the circumstances of the work, including the amount of the substance involved;

(e) activities, such as maintenance, where there is the potential for a high level of exposure;

(f) any relevant workplace exposure limit or similar occupational exposure limit;

(g) the effect of preventive and control measures which have been or will be taken in accordance with regulation 7;

(h) the results of relevant health surveillance;

(i) the results of monitoring of exposure in accordance with regulation 10;

(j) in circumstances where the work will involve exposure to more than one substance hazardous to health, the risk presented by exposure to such substances in combination;

(k) the approved classification of any biological agent; and

(l) such additional information as the employer may need in order to complete the risk assessment.

(3) The risk assessment shall be reviewed regularly and forthwith if–

(a) there is reason to suspect that the risk assessment is no longer valid;

(b) there has been a significant change in the work to which the risk assessment relates; or

(c) the results of any monitoring carried out in accordance with regulation 10 show it to be necessary, and where, as a result of the review, changes to the risk assessment are required, those changes shall be made.

(4) Where the employer employs 5 or more employees, he shall record–

(a) the significant findings of the risk assessment as soon as practicable after the risk assessment is made; and

(b) the steps which he has taken to meet the requirements of regulation 7.

In making assessments, it should be noted that *Regulation 3(1)* also requires employees to take account, so far as is reasonably practicable, of others who may be affected by the work, e.g. visitors, contractors, customers, passers-by and members of the emergency services.

Essentially there are four aspects to be considered during the assessment:

1 Identification of risks to the health of employees or others.

2 Consideration of whether it is reasonably practicable to prevent exposure to hazardous substances creating risks (by elimination or substitution).

3 If prevention is not reasonably practicable, identification of the measures necessary to achieve adequate control of exposure, as required by *Regulation* 7 (these control measures are described in greater detail later in the chapter).

4 Identification of other measures necessary to comply with *Regulations 8* to 13

- use of control measures;
- maintenance, examination and test of control measures, etc.;
- monitoring of exposure at the workplace;
- health surveillance;
- provision of information, instruction and training;
- arrangements for accidents, incidents and emergencies

(All of these are described more fully later in the chapter.)

The assessment must be 'suitable and sufficient', and its complexity will vary considerably dependent on the circumstances. In some cases it will only be necessary to study suppliers' data sheets and decide whether existing practices are sufficient to achieve adequate control. However, for more complex and variable processes with greater numbers of hazardous substances involved, the assessment will need to be much more detailed.

This is demonstrated by the sample assessment records provided later in the chapter. A COSHH assessment in an office environment will be a straightforward matter, not requiring any specialist skills. However, an assessment in a manufacturing plant handling large quantities of chemicals will be a much more complex affair.

The planning and preparation for a COSHH assessment should follow a similar pattern to that described in **CHAPTER 4: CARRYING OUT RISK ASSESSMENTS** for general risk assessments. Some of those principles are repeated below and additional guidance specific to COSHH assessments is also provided.

Who will carry out the assessments?

9.7 As described in **CHAPTER 4: CARRYING OUT RISK ASSESSMENTS**, *Regulation* 7 of the *Management Regulations 1999* requires employers to appoint a competent person or persons to assist them in complying with their duties – which include carrying out COSHH assessments. *Regulation 12(4)* of the *COSHH Regulations* themselves also states:

> Every employer shall ensure that any person (whether or not his employee) who carries out any work in connection with the employer's duties under these Regulations has suitable and sufficient information, instruction and training.

As with general risk assessments the degree of knowledge and experience required will depend upon the circumstances. An assessment in a workplace containing a few straightforward uses of hazardous substances may be carried out by someone without any specialist qualifications or experience but with the capability to understand and apply the principles contained in this chapter. (However, such a person may need access to someone with greater health and safety knowledge and a more detailed understanding of the *COSHH Regulations*.) Qualified and experienced individuals may be capable of carrying out detailed assessments of relatively high-risk situations themselves, although they may well need to consult others during the assessment process.

Higher risk situations with much more complex and significant use of hazardous substances are more likely to require multi-disciplinary teams, some of whom have specialist knowledge and experience. Possible candidates for such a team might be:

- health and safety officers or managers;
- occupational hygienists;
- occupational health nurses;
- physicians with occupational health experience;
- chemical or process engineers;
- maintenance or ventilation engineers.

As with general risk assessments, the combination of an 'insider' (a manager, supervisor or engineer from the department concerned) working together with an 'outsider' (a health and safety specialist) is often beneficial. Someone should be appointed to co-ordinate the work of such a team.

In some situations an in-house team or an individual with little or no specialist qualification or experience may be able to carry out the majority of its COSHH assessments. They can then identify those areas where more specialised assistance is required to come to a conclusion, or where occupational hygiene surveys are necessary before a decision can be made.

How will the assessments be organised?

9.8 Larger workplaces will usually need to be divided into manageable assessment units which might be based on:

- departments or sections;
- buildings or rooms;
- process lines;
- activities or services.

Where an assessment team is used, those individuals most suited for a particular situation can be selected to assess that unit. As an example, a small paint manufacturing plant might be divided into units as follows:

- Raw material storage – including any tank farms and drum storage areas.
- Manufacturing – possibly split into different product lines or buildings.
- Container filling – if carried out separately from process lines.
- Product storage – warehouses and container storage areas.
- Maintenance – including maintenance activities in process areas and use of hazardous substances in maintenance work.
- Administration/miscellaneous.

Gathering information

9.9

- **Identification of substances present**
  All hazardous substances must be taken into account during the assessments and as many of these as possible should be identified before the assessment takes place, e.g.:
  - Raw materials and the contents of stores.
  - Substances produced by processes:
    - intermediate compounds
    - products
    - waste and by-products
    - emissions from the process.
  - Substances used in maintenance and cleaning work.
  - Buildings and the work environment:
    - surface treatments
    - pollution or contaminants
    - bird droppings, animal faeces, etc.
    - possible legionella sources.
  - Substances involved in the activities of others, e.g. contractors.
- **Hazards associated with substances**
  Information should be gathered on the hazardous substances identified. Potential sources include:
  - Manufacturers' or suppliers' safety data sheets.
  - Information provided on containers or packaging (including the symbols required by the *CHIP Regulations*).

- HSE publications (**REFS 2–28** indicate the range of information available – the HSE catalogue includes far more).
- Reference books and technical literature.
- Direct enquiries to specialists or suppliers.

This information should primarily concern the hazards associated with the substance but note should also be taken of recommendations on control methods. However, caution should be exercised in this latter respect, particularly in relation to information from suppliers. The substance's actual circumstances of use may constitute a greater or a lesser degree of risk than the supplier anticipated – it is for the assessor(s) to determine what control measures are necessary in each situation. Further enquiries to suppliers may be necessary in some cases.

- **Previous assessments, surveys, etc.**
  Even if previous COSHH assessments were inadequate or are now out of date, they may still contain information of value in the assessment process. The results of previous occupational hygiene surveys, e.g. dust, gas or vapour concentrations, are also likely to be useful as will the collective results of any health surveillance work which has previously been carried out.
- **Existing control measures**
  Existing control measures may be described in operating procedures, health and safety rule books or listings of PPE requirements or may be referred to within training programmes. In some cases it may be appropriate to seek detailed specification information in relation to control measures, e.g. design flow rates for ventilation equipment or performance standards for PPE such as respiratory protection or gloves. (Some of this information may be sought during the assessment itself or prior to the preparation of the assessment record.)

Prevention or control of exposure

9.10 Before commencing the assessment itself, it is important to have an understanding of what the regulations require in terms of prevention or control and also the range of control measures available.

Hierarchy of measures

9.11 *Regulation* 7 of the *COSHH Regulations* sets out a clear hierarchy of measures which must be taken to prevent or control exposure to hazardous substances. It states:

(1) Every employer shall ensure that the exposure of his employees to substances hazardous to health is either prevented or, where this is not reasonably practicable, adequately controlled.

(2) In complying with his duty of prevention under paragraph (1), substitution shall by preference be undertaken, whereby the employer shall avoid, so far as is reasonably practicable, the use of a substance hazardous to health at the workplace by replacing it with a substance or process which, under the conditions of its use, either eliminates or reduces the risk to the health of the employees.

(3) Where it is not reasonably practicable to prevent exposure to a substance hazardous to health, the employer shall comply with his duty of control under paragraph (1) by applying protection measures appropriate to the activity and consistent with the risk assessment, including in order of priority–

(a) the design and use of appropriate work processes, systems and engineering controls and the provision and use of suitable work equipment and materials;

(b) the control of exposure at source, including adequate ventilation systems and appropriate organisational measures; and

(c) where adequate control of exposure cannot be achieved by other means, the provision of suitable personal protective equipment in addition to the measures required by sub-paragraphs (a) and (b).

Thus a clear order of preference is established:

- prevention of exposure (by elimination or substitution);
- appropriate work processes, systems and engineering controls (e.g. enclosure) and suitable work equipment and materials;
- adequate control at source (by ventilation or organisational measures);
- adequate control through PPE.

(*Sub-paragraph (4)* of *Regulation* 7 contains further details of what the measures referred to in *Paragraph (3)* must include – these are referred to later in the chapter.)

Prevention of exposure

9.12 The first preference should always be to prevent exposure to hazardous substances.

Means available include:

- Changing work methods, e.g. cleaning items using ultrasonic techniques or high pressure water jets rather than with solvents.
- Using non-hazardous (or less hazardous) alternatives, e.g. replacing solvent-based paints or inks by water-based ones (or ones utilising less-hazardous solvents).

- Using less-hazardous forms of substances, e.g. substituting powdered materials by granules or pellets, using hazardous substance in a more dilute form.
- Modifying processes, e.g. eliminating production of hazardous by-products, waste or emissions by altering process parameters such as temperature or pressure.

Exposure must be prevented 'so far as is reasonably practicable' – a definition of this qualifying phrase is provided in **CHAPTER 1: INTRODUCTION**. The level of risk will be determined by the type and severity of the hazards associated with a substance and its circumstances of use.

Factors weighing against prevention of exposure being reasonably practicable might be the costs associated with using alternative methods or materials, the detrimental effect of alternatives on the product or additional risks introduced by the alternatives.

For example:

- ultrasonic cleaning may not achieve the desired results;
- use of high pressure water can introduce additional risks;
- water-based paints may increase problems of corrosion;
- alternative solvents may be more highly flammable.

Nevertheless this option can often prove reasonably practicable – much progress has been made in recent years in the substitution of hazardous materials. The HSE booklet 'Seven steps to successful substitution of hazardous substances' (**REF. 7**) provides useful guidance.

Control measures, other than PPE

9.13 *Paragraph (4)* of *Regulation* 7 of the *COSHH Regulations* states that the control measures required by *Paragraph (3)* shall include:

(a) arrangements for the safe handling, storage and transport of substances hazardous to health, and of waste containing such substances, at the workplace;

(b) the adoption of suitable maintenance procedures;

(c) reducing, to the minimum required for the work concerned–

  (i) the number of employees subject to exposure

  (ii) the level and duration of exposure, and

  (iii) the quantity of substances hazardous to health present at the workplace;

(d) the control of the working environment, including appropriate general ventilation; and

(e) appropriate hygiene measures including adequate washing facilities.

Examples of measures which might be necessary include:

- Enclosed process vessels.
- Enclosed transfer systems, e.g. use of pumps or conveyors, rather than open transfer.
- Use of closed and clearly labelled containers.
- Plant, processes and systems of work which minimise the generation of, or suppress and contain, spills, leaks, dust, fumes and vapours. (This often involves a combination of partial enclosure and the use of LEV. The design of LEV is a subject in its own right – see **REF. 29**.)
- Adequate levels of general ventilation.
- Suitable maintenance procedures.
- Limitation of the quantities of hazardous substances in workplaces.
- Minimising the numbers of persons in areas where hazardous substances are used.
- Prohibition of eating, drinking and smoking in areas of potential contamination.
- Provision of showers in addition to normal washing facilities.
- Regular cleaning of walls and surfaces.
- Use of signs to indicate areas of potential contamination.
- Safe storage, handling and disposal of hazardous substances (including hazardous waste).

Paragraphs (5) and *(6)* of *Regulation* 7 set out measures required to control exposure to carcinogens and biological agents – these are dealt with later in the chapter.

Control using PPE

9.14 Control using PPE should be the final option, where prevention or adequate control of exposure by other means are not reasonably practicable. This might be the case where:

- the scale of use of hazardous substances is very small;
- adequate control by other means is not technically feasible;
- control by other means is excessively expensive or difficult in relation to the level of risk;

- PPE is utilised as a temporary measure pending the implementation of adequate control by other means;
- employees are dealing with emergency situations;
- infrequent maintenance activities are carried out.

The PPE necessary to control risks from hazardous substances might be:

- respiratory protective equipment (RPE) – see **REF. 30**
- protective clothing;
- hand or arm protection (usually gloves) – see **REFS 31** and **32**;
- eye protection;
- protective footwear.

Paragraph (9) of *Regulation* 7 of the *COSHH Regulations* states that:

> Personal Protective Equipment provided by an employer in accordance with this regulation shall be suitable for the purpose and shall–
>
> (a) comply with any provision in the Personal Protective Equipment Regulations 2002 which is applicable to that item of personal protective equipment; or
>
> (b) in the case of respiratory protective equipment, where no provision referred to in sub-paragraph (a) applies, be of a type approved or shall conform to a standard approved, in either case, by the Executive.

The HSE has detailed guidance available on PPE generally (**REF. 33**) and RPE in particular (**REF. 30**). Further guidance on the use of PPE and the assessment of PPE requirements is provided in **CHAPTER 13**: **ASSESSMENT OF PERSONAL PROTECTIVE EQUIPMENT (PPE) REQUIREMENTS.**

Adequate control

9.15 *Paragraphs (7)* and *(11)* of *Regulation* 7 of the *COSHH Regulations* define what is meant by the term 'adequate control' in respect of risks from hazardous substances:

> (7) Without prejudice to the generality of paragraph (1), where there is exposure to a substance hazardous to health, control of that exposure shall only be treated as being adequate if –
>
> (a) the principles of good practice for the control of exposure to substances hazardous to health set out in Schedule 2A are applied;
>
> (b) any workplace exposure limit approved for the substance is not exceeded; and

(c) for a substance –

(i) which carries the risk phrase R45, R46 or R49, or for a substance or process which is listed in Schedule 1; or

(ii) which carries the risk phrase R42 or R42/43, or which is listed in section C of HSE publication "Asthmagen? Critical assessments of the evidence for agents implicated in occupational asthma" as updated from time to time, or any other substance which the risk assessment has shown to be potential cause of occupational asthma, exposure is reduced to as low as is reasonably practicable.

(11) In this regulation, "adequate" means adequate having regard only to the nature of the substance and the nature and degree of exposure to substances hazardous to health and "adequately" shall be construed accordingly.

The principles of good practice set out in *Schedule 2A* largely duplicate and expand on the requirements of the COSHH Regulations, although it should be noted that they refer to controlling exposure 'by measures that are proportionate to the health risk'. Workplace exposure limits (WELs) replace the terms previously used – 'maximum exposure limits' and 'occupational exposure standards'. Listings of WELs are published regularly by the HSE (**REF. 34**). It should be noted that for the substances referred to in sub-paragraph (c) above, exposure must be reduced to as low a level as is reasonably practicable.

In some cases it will be relatively easy to make a judgement as to whether exposure is controlled to levels well within the WEL, but in other situations it may be necessary to carry out an occupational hygiene survey or even arrange for periodic or ongoing monitoring of exposure. The techniques for doing this are described later in the chapter.

Carcinogens, mutagens and biological agents

9.16 *Paragraphs (5)* and *(6)* of *Regulation* 7 of the *COSHH Regulations* set out a number of specific measures which must be applied where it is not reasonably practicable to prevent exposure to carcinogens, mutagens and biological agents. (These are in addition to the measures required by *Paragraph (3)* of the Regulation.)

Measures specified for controlling carcinogens and mutagens are:

- totally enclosing process and handling systems (unless not reasonably practicable);
- prohibition of eating, drinking and smoking in areas that might be contaminated;
- cleaning of floors, walls and other surfaces regularly and whenever necessary;
- designating areas of potential contamination by warning signs;
- storing, handling and disposing of carcinogens or mutagens safely, using closed and clearly labelled containers.

Appendix 1 of the HSE COSHH ACOP booklet (**REF. 1**) provides further detail on the control of carcinogenic and mutagenic substances whilst Annex 1 of the booklet contains a background note on occupational cancer.

Measures specified for controlling biological agents are:

- use of warning signs (a biohazard sign is shown in *Schedule 3* of the Regulations);
- specifying appropriate decontamination and disinfection procedures;
- safe collection, storage and disposal of contaminated waste;
- testing for biological agents outside primary confinement areas;
- specifying work and transportation procedures;
- where appropriate, making available effective vaccines;
- instituting appropriate hygiene measures;
- control and containment measures where human patients or animals are known (or suspected to be) infected with a Group 3 or 4 biological agent.

Paragraph (10) of *Regulation* 7 requires work with biological agents to be in accordance with *Schedule 3* to the Regulations which is contained in the HSE COSHH ACOP booklet (**REF. 1**). This sets out several detailed requirements, including standards for containment measures.

Making the assessment

9.17 As was the case for general risk assessments, time spent in making COSHH assessments can both be reduced and made more productive by good planning and preparation. Nevertheless, observations of workplaces and work activities must still be an integral part of the COSHH assessment process.

Observations

9.18 Sufficient observations must be made to arrive at a conclusion on the adequacy of control measures for exposures to all the hazardous substances in the workplace – be they raw materials, products, by-products, waste, emissions, etc. This should include the manner and the extent to which employees (and others) are exposed and an evaluation of the equipment and working practices intended to achieve their control.

Aspects to be considered include:

- Are specified procedures being followed?
- Does LEV or general ventilation appear effective?
- Is PPE being used correctly?
- Is there evidence of leakage, spillage or dust accumulation?

- What equipment is available for cleaning?
- Are dusts, fumes or strong smells evident in the atmosphere?
- Are there any restrictions relating to eating, drinking or smoking?
- What arrangements are there for storage, cleaning or maintenance of PPE?
- Are there suitable arrangements for washing, showering or changing clothing?
- Are special arrangements for laundering clothing required?
- What arrangements are in place for storage and disposal of waste?
- Is there potential for major spillages or leaks?
- What emergency containment equipment or PPE is available?

Obviously this list could be extended significantly and some of the aspects could be broken down into further sub-components.

Discussions

9.19 It is also important to talk to those working with hazardous substances, their safety representatives and those managing or supervising their work. Assessors' questions might relate to working practices, awareness of risks or precautions, the effectiveness of precautions or experience of problems. Questions might include:

- Why is the task done that way?
- Are the present work practices typical?
- How effective is the LEV/general ventilation? What happens if it ever breaks down?
- What types of work require PPE to be used?
- Are there any problems with the PPE?
- Have workers experienced any health problems?
- How often is the workplace cleaned?
- What methods are used for cleaning?
- How is waste disposed of?
- Is there any possibility of leaks or emergencies arising?
- What would happen in such a case?
- What are the arrangements for washing/cleaning/maintaining PPE?

Once again the potential list of questions is endless and there can be many variations on the above themes.

Sources of help

9.20 Further guidance on the COSHH process is available in the HSE publication *COSHH essentials* (**REF. 35**) which was developed by the HSE's Advisory Committee on Toxic Substances. It provides generic risk assessments for a wide range of substances covered by CHIP and COSHH. It can also be accessed electronically via the HSE website: www.hse.gov.uk. Control advice is provided for common tasks, e.g. mixing, weighing, spray painting.

Further tests and investigations

9.21 Assessment in the workplace may reveal the need for further tests on the effectiveness of control measures before the assessment can be concluded. The tests may be in the form of occupational hygiene surveys – usually to measure the airborne levels of dust, fume, vapour or gas and compare them with the relevant WEL. Techniques available for such survey work are described later in this chapter. Alternatively there may be a need to measure the effectiveness of LEV or general ventilation – this too is referred to later in the chapter.

Further investigations may also be necessary into matters such as:

- hazards associated with substances not previously identified during the preparatory phase of the assessment;
- the specifications for PPE found to be in use during the course of the assessment;
- the feasibility of alternative work methods (e.g. preventing exposure) or alternative methods of control to those observed.

Preparation of assessment records

9.22 The guidance on note taking and the preparation of assessment records contained in **CHAPTER 4: CARRYING OUT RISK ASSESSMENTS** is equally relevant to COSHH assessments. Several examples of completed assessment records are provided on the succeeding pages, although it should be stressed that (as for general risk assessments) no single record format will automatically cater for all types of workplaces or work activities. The essential components which must be included in COSHH assessment records are:

- the work activities involving risks from hazardous substances;
- information as to the hazardous substances involved (and their form, e.g. liquid, powder, dust, etc.);
- control measures which are (or should be) in place;
- improvements identified as being necessary.

As for general risk assessments, employers with less than five employees are exempt from the requirement to record assessments, although the COSHH ACOP recommends recording the significant findings in any case. In all other cases the significant findings must be recorded and kept readily accessible to those who may need to know the results. Employees or their representatives should be informed of the results of COSHH assessments.

Sample assessment records

9.23

ACORN ESTATE AGENTS, NEWTOWN

RISK ASSESSMENT

REFERENCE NUMBER: 6	RISK TOPIC/ISSUE: HAZARDOUS SUBSTANCES	SHEET 1 of 1
CROSS REFERENCES:		
RISKS IDENTIFIED	PRECAUTIONS IN PLACE	RECOMMENDED IMPROVEMENTS
The substances listed below could present risks to both staff and clients. OFFICE MATERIALS *With warning symbols* Old correction fluid (Harmful) New correction fluid (Flammable) Spray adhesive (Harmful) *Without warning symbols* Various felt tip pens (solvent based) Photocopier and printer toners	Office supplies are kept in a cupboard in a part of the office not normally accessible to clients. Generally substances are only used in very small quantities for short periods presenting no significant risk. If the spray adhesive is used for more than a couple of minutes, a nearby window is opened which provides adequate ventilation. The photocopier is used in well-ventilated areas and there are no noticeable ozone smells, even on long copying runs.	Provide disposable gloves for cleaning significant spillages of photocopier or printer toner.
CLEANING MATERIALS *With warning symbols* Thick bleach (Irritant) Polish stripper (Irritant) Acid descaler (Corrosive) *Without warning symbols* Furniture polish Floor polish Window and glass cleaner	The cleaner's cupboard is kept locked except when substances are being removed. Suitable gloves are provided (and worn) for handling the irritant and corrosive substances in concentrated form. The cleaner is aware of the risks of mixing bleach with other substances, e.g. the acid descaler.	Investigate replacing the thick bleach by a more dilute solution. Ensure the relief cleaner is also made aware of these risks.
SIGNATURE(S): *K Stephenson, R Lewis*	NAME(S): *K Stephenson, R Lewis*	DATE *4/11/98*
DATES FOR RECOMMENDATION FOLLOW UP: *December 1998*		NEXT ROUTINE REVIEW: *November 2003*

TIMREK Engineering

COSHH ASSESSMENT

Main risks

1. Electric arc welding, mainly of large-and medium-sized structures.
2. Oxy-acetylene burning, welding and brazing, as above.
3. Cleaning of parts in proprietary unit (using paraffin based solvent).
4. Use of cutting oils on the lathe and milling machine.
5. Application of solvent-based primers by brush (not spray). (Some primers are designated as 'harmful'.)
6. Small scale use of a variety of cleaning solvents, lubricants, adhesives and paints using aerosol sprays or brush application. (Several of these are labelled as harmful, irritant or corrosive.)

Control measures

1. & 2. Limited work carried out – never more than five minutes in duration or a total of 20 minutes per day. Workshop door normally open. Portable extraction unit available but currently broken.

3. Unit includes brush applicator. Suitable gauntlets provided and used. Solvent changed regularly by suppliers.

4. Splash guards in place. Suitable gloves provided and used for handling oils and swarf. General ventilation adequate to remove any fume or mist. Employees exposed report no breathing or skin problems.

5. Never more than 15 minutes work per day. Priming done near open workshop door if possible but sometimes there is a strong smell. Suitable gloves are provided and worn.

6. Substances only used for extremely short periods of time. Gloves and eye protection (safety spectacles) provided.

Recommendations

1. & 2.

   a) Arrange for portable extraction unit to be repaired.

   b) Carry out weekly visual check on extraction unit.

   c) Arrange for insurance company engineer to conduct examination and test every 14 months.

3. None.

4. Check annually with staff that they are not experiencing problems (make diary note).

5. Investigate the feasibility of using water-based primer.

6. a) Prepare a table showing which substances require the use of gloves and/or safety spectacles.

   b) Display the table on the notice-board and draw it to the attention of all employees.

Signed: *A. F. Rogerson* Date: *4.08.2000*

Follow up of recommendations: *October 2000 (make diary note)*

Review assessment: August 2002

RAINBOW PRODUCTS

COSHH ASSESSMENT

ASSESSMENT UNIT	Mixing Hall
ACTIVITIES	Manufacture of solvent-based paints and other surface treatments. Solvents are pumped into the mixing vessels from an external tank farm. Solid constituents and some liquid components are charged into the mixing vessels from platforms above. After mixing, the products are pumped directly to filling stations in a neighbouring building.
SUBSTANCES USED (Data sheet file references)	4, 7, 14, 23, 49, 50, 51, 60, 73, 80, 92, 106, 117. Detailed formulations for each product are available from the Quality Control Department.
MAIN RISKS	Dust – from solid constituents during charging Solvent vapours escaping – from the charging hatches – from minor leaks at valves and pipe connections – from liquid components during charging Entry into mixing vessels for cleaning or maintenance purposes.
THOSE AT RISK	Mixing Hall production employees. Maintenance staff when working in the area. Contractors involved in vessel cleaning or maintenance work.
CONTROLS AND OTHER PRECAUTIONS IN PLACE	Good general ventilation (specification 20 air changes per hour). Annual surveys of solvent vapours show levels well below all WELs. Hoods with LEV over each vessel charging position. These are examined and tested annually by the Maintenance Department. Exposure Monitoring Surveys (August 2002, May 2005) show production employees as well within WELs for all dusts and solvents. Disposable dust masks are available for vessel charging, although their use is not compulsory. A portable vacuum cleaning unit with suitable filter is available when required. Maintenance carry out an annual physical inspection of all mixer vessels and pipelines. A weigh station is provided in a fume cupboard for weighing out smaller quantities of solid constituents (this also is examined and tested annually). Entry into vessels is controlled by the permit to work system. All employees are subject to the company's annual health screening programme.
OBSERVATIONS	Discarded empty paper sacks were strewn on several loading platforms. Liquid component transfer containers had been left open on platforms 1 and 4 (still containing residual materials). The ventilation at charging stations 2 and 4 appeared inadequate. There were dust accumulations on all loading platforms. A brush appears to have been used to sweep up dust on some platforms. A pipe flange below mixer 3 had developed a small leak.

RECOMMENDATIONS

1. Improve the sack disposal containers on all loading platforms.
2. Remind staff of the importance of disposing of sacks correctly.
3. Remind staff that liquid transfer containers should have their lids replaced after use.
4. Rectify the ventilation at charging stations 2 and 4.
5. Introduce simply weekly checks on the ventilation at all charging stations.
6. Introduce weekly cleaning for all loading platforms.
7. Remind staff of the importance of using vacuum methods for cleaning.
8. Investigate obtaining vacuum units which can be lifted onto the platforms more easily.
9. Repair the leaking flange below mixer 3.
10. Increase the frequency of pipeline inspections to six monthly.
11. Develop a standard procedure for entering vessels for cleaning or maintenance purposes (linked to the permit to work system).
12. Investigate alternative cleaning methods avoiding the need for entry, e.g. immersion in solvent over weekend periods.

Signatures: *A Storm, P Gold* Date: *2 July 2005*

Recommendation follow up: *October 2005*

Assessment Review: *July 2007*

After the assessment

Review and implementation of recommendations

9.24 The guidance provided in **CHAPTER 4: CARRYING OUT RISK ASSESSMENTS** on the review of recommendations resulting from risk assessments and the implementation of an action plan is equally applicable to assessments made under the *COSHH Regulations*. Some recommendations could relate to the ongoing maintenance of effective control measures and many of these aspects are covered by further specific requirements of the *COSHH Regulations*.

Use of control measures

9.25 *Regulation 8* of the *COSHH Regulations* places duties on both employers and employees in respect of the proper use of control measures, including PPE.

> (1) Every employer who provides any control measure, other thing or facility in accordance with these Regulations shall take all reasonable steps to ensure that it is properly used or applied as the case may be.
>
> (2) Every employee shall make full and proper use of any control measure, other thing or facility provided in accordance with these Regulations and, where relevant shall—
>
> > (a) take all reasonable steps to ensure it is returned after use to any accommodation provided for it; and
> >
> > (b) if he discovers any defect therein, report it forthwith to his employer.

As far as the employer is concerned, this should form part of the monitoring stage in the management cycle described in **CHAPTER 8: IMPLEMENTATION OF PRECAUTIONS**. Aspects where monitoring is likely to be particularly relevant are:

- correct use of LEV equipment;
- compliance with specified systems of work;
- compliance with PPE requirements;
- storage and maintenance of PPE;
- compliance with requirements relating to eating, drinking or smoking;
- condition of washing and showering facilities and personal hygiene standards;
- whether defects are being reported by employees.

Where employees are unwilling to use control measures properly, employers should consider the use of disciplinary action, particularly in relation to persistent offenders.

Maintenance, examination and testing of control measures

9.26 *Regulation 9* of the *COSHH Regulations* contains both general and specific requirements for the maintenance of control measures:

(1) Every employer who provides any control measure to meet the requirements of regulation 7 shall ensure that –

(a) in the case of plant and equipment, including engineering controls and personal protective equipment, it is maintained in an efficient state, in efficient working order, in good repair and in a clean condition; and

(b) in the case of the provision of systems of work and supervision and of any other measure, it is reviewed at suitable intervals and revised if necessary.

(2) Where engineering controls are provided to meet the requirements of regulation 7, the employer shall ensure that thorough examination and testing of those controls is carried out —

(a) in the case of local exhaust ventilation plant, at least once every 14 months, or for local exhaust ventilation plant used in conjunction with a process specified in Column 1 of Schedule 4, at not more that the interval specified in the corresponding entry in Column 2 of that Schedule; or

(b) in any other case, at suitable intervals.

(3) Where respiratory protective equipment (other than disposable respiratory protective equipment) is provided to meet the requirements of regulation 7, the employer shall ensure that thorough examination and, where appropriate, testing of that equipment is carried out at suitable intervals.

(4) Every employer shall keep a suitable record of the examinations and tests carried out in pursuance of paragraphs (2) and (3) and of repairs carried out as a result of those examinations and tests, and that record or a suitable summary thereof shall be kept available for at least 5 years from the date on which it was made.

Paragraphs (5), (6) and *(7)* contain requirements specifically relating to PPE – these are examined later.

General maintenance of controls

9.27 The COSHH ACOP states that, where possible, all engineering control measures should receive a visual check at appropriate intervals, and in the case of LEV and work enclosures, at least once every week. Such checks may simply confirm that there are no apparent leaks from vessels or pipes and that LEV or cleaning equipment appear to be in working order. No records of such checks need to be kept, although it is good practice (and prudent) to do so.

Paragraph (2) of *Regulation 9* of the *COSHH Regulations* requires thorough examinations and tests of engineering controls. Requirements relating to LEV are reviewed below but for other engineering controls such examinations and test must be 'at suitable intervals', and suitable records must be kept for at least five years. The nature of examinations and tests will depend upon the engineering control involved and the potential consequences of its deterioration or failure. Examples might involve:

- detailed visual inspections of tanks and pipelines;
- inspections and non-destructive testing of critical process vessels;
- testing of detectors and alarm systems;
- planned maintenance of general ventilation equipment;
- checks on filters in vacuum cleaning equipment.

Persons carrying out maintenance, examinations and testing must be competent for the purpose in accordance with *Regulation 12(4)* of the *COSHH Regulations*.

Local Exhaust Ventilation (LEV) plant

9.28 Most LEV systems must be thoroughly examined and tested at least once every 14 months, although *Schedule 4* to the *COSHH Regulations* requires increased frequencies for LEV used in conjunction with a handful of specified processes. Dependent upon the design and purpose of the LEV concerned, the examination and test might involve one or more of the following:

- visual inspection of the LEV equipment;
- air flow measurements using an air velocity meter (involving comparisons with recommended capture and duct velocities);
- static pressure measurements (and comparison with design or commissioning pressures);
- visual checks of efficiency using smoke generators or dust lamps;
- air sampling to confirm efficiency levels (applying the sampling methods described later in this chapter);
- filter integrity tests to confirm filter efficiency;
- checks on air flow sensors.

Further guidance is available in an HSE booklet (**REF. 36**), and details of the records which should be kept in relation to those examinations and tests are contained in the COSHH ACOP (**REF. 1**).

Personal Protective Equipment

9.29 All types of PPE are subject to the general maintenance requirements contained in *Paragraph (1)* of *Regulation 9* of the *COSHH Regulations* whilst *Paragraph (3)* contains specific requirements relating to non-disposable RPE.

Further requirements in respect of PPE are contained in *Paragraphs (5), (6)* and *(7)* of *Regulation 9* which state:

(5) Every employer shall ensure that personal protective equipment, including protective clothing, is:

- (a) properly stored in a well-defined place;
- (b) checked at suitable intervals; and
- (c) when discovered to be defective, repaired or replaced before further use.

(6) Personal protective equipment which may be contaminated by a substance hazardous to health shall be removed on leaving the working area and kept apart from uncontaminated clothing and equipment.

(7) The employer shall ensure that the equipment referred to in paragraph (6) is subsequently decontaminated and cleaned or, if necessary, destroyed.

Some types of PPE can easily be seen to be defective by the user whilst in other cases (e.g. for gloves or clothing providing protection against strongly corrosive chemicals) it may be appropriate to introduce more formalised inspection systems.

For non-disposable RPE the COSHH ACOP states that thorough examinations and, where appropriate, tests should be made *at least* once every month, although it suggests that for RPE used less frequently, periods up to three months are acceptable. Alternatively, examination and testing prior to next use may be more appropriate. Emergency escape-type RPE should be examined and tested in accordance with the manufacturer's instructions.

For simple respirators a visual examination of the condition of the facepiece, straps, filters and valves is sufficient. In the case of powered and power-assisted respirators tests should be made on the condition of the battery pack and the adequacy of the flow rate. However, for airline-fed RPE the quality and flow of the air supply should also be tested. For RPE supplied from compressed gas cylinders more detailed examinations and testing will be required, including a check on the pressure in the cylinders. Details of the records which should be kept are contained in the COSHH ACOP (**REF. 1**). Further guidance is available in an HSE booklet (**REF. 30**).

Monitoring exposure at the workplace

9.30 Reference was made earlier in the chapter to the possible need to carry out air testing in the workplace as part of the COSHH assessment process in order to determine the adequacy of control measures. Such testing may also be necessary in order to ensure that adequate control continues to be maintained and this is a requirement of *Regulation 10* of the *COSHH Regulations*:

(1) Where the risk assessment indicates that —

(a) it is requisite for ensuring the maintenance of adequate control of the exposure of employees to substances hazardous to health; or

(b) it is otherwise requisite for protecting the health of employees, the employer shall ensure that the exposure of employees to substances hazardous to health is monitored in accordance with a suitable procedure.

(2) Paragraph (1) shall not apply where the employer is able to demonstrate by another method of evaluation that the requirements of regulation 7(1) have been complied with.

(3) The monitoring referred to in paragraph (1) shall take place —

(a) at regular intervals; and

(b) when any change occurs which may affect that exposure.

(4) Where a substance or process is specified in Column 1 of Schedule 5, monitoring shall be carried out at least at the frequency specified in the corresponding entry in Column 2 of that Schedule.

(5) The employer shall ensure that a suitable record of any monitoring carried out for the purpose of this regulation is made and maintained and that record or a suitable summary thereof is kept available -

(a) where the record is representative of the personal exposures of identifiable employees, for at least 40 years; or

(b) in any other case, for at least 5 years, from the date of the last entry made in it.

(6) Where an employee is required by regulation 11 to be under health surveillance, an individual record of any monitoring carried out in accordance with this regulation shall be made, maintained and kept in respect of that employee.

(7) The employer shall—

(a) on reasonable notice being given, allow an employee access to his personal monitoring record;

(b) provide the Executive with copies of such monitoring records as the Executive may require; and

(c) if he ceases to trade, notify the Executive forthwith in writing and make available to the Executive all monitoring records kept by him.

Schedule 5 to the *COSHH Regulations* automatically requires monitoring to be carried out in processes using vinyl chloride monomer and electrolytic chromium plating.

The range of equipment available for monitoring air quality in the workplace is constantly expanding but there are three main types:

1. Chemical indicator tubes.
2. Direct reading instruments.
3. Sampling pumps and filter heads.

Chemical indicator tubes

9.31 Tubes are available from a variety of suppliers to measure a wide range of airborne substances. A measured volume of air is drawn through the tube, usually using a hand bellows or a mechanical pump. The contaminant (usually a gas, vapour or aerosol) reacts with the chemicals contained in the tube to produce a colour change. The concentration is measured by calibrated markings showing how far the colour change has penetrated into the tube or by comparing the intensity of the colour change with a calibrated chart. Different types of tubes can measure concentrations in both the short term and the long term.

Direct reading instruments

9.32 Instruments can provide direct measurements of the concentrations of gases, vapours or dust present in the atmosphere. Some instruments are portable (particularly useful for measuring contaminants in confined spaces) whereas others are fixed, sometimes as part of detection networks. As well as providing instant measurement of concentrations, instruments can be set to trigger off alarms when specified concentrations are reached, e.g. 50% of the WEL.

Sampling pumps and filter heads

9.33 Small battery-operated pumps can be used to draw air through filter heads or some absorbent medium. The samples collected can then be weighed, chemically analysed or counted under microscopes (e.g. in the case of asbestos fibres). The concentration can be determined from the weight of the sample (or the number of fibres counted) and the volume of air drawn through by the pump. Different filtration and absorbent materials are available to sample a wide range of contaminants.

Interpretation of monitoring results

9.34 Most types of monitoring equipment can be used either to monitor the concentration of a substance in a given part of a workplace or to measure the concentration in the breathing zone of an individual employee, to provide an indication of their exposure. Care should be taken in interpreting all survey results – the conditions measured may not be representative of those normally encountered and allowances should be made for the accuracy of the sampling method. Some occupational hygienists deliberately sample what they believe to be the worst conditions first. If these circumstances produce concentrations which are comfortably within the relevant WEL, then the need for further monitoring may be greatly reduced or eliminated. Note should be taken of the requirement to retain all exposure monitoring records for five years and those relating to the personal exposure of identifiable individuals for at least 40 years. The HSE provide considerable technical guidance on monitoring methods, particularly in the HSE series of *Methods for the Determination of Hazardous Substances* publications (MDHS).

Health surveillance

9.35 *Paragraphs (1)* and *(2)* of *Regulation 11* of the *COSHH Regulations* contain the main requirements relating to the need for health surveillance:

(1) Where it is appropriate for the protection of the health of his employees who are, or are liable to be, exposed to a substance hazardous to health, the employer shall ensure that such employees are under suitable health surveillance.

(2) Health surveillance shall be treated as being appropriate where —

(a) the employee is exposed to one of the substances specified in Column 1 of Schedule 6 and is engaged in a process specified in Column 2 of that Schedule, and there is a reasonable likelihood that an identifiable disease or adverse health effect will result from that exposure; or

(b) the exposure of the employee to a substance hazardous to health is such that—

(i) an identifiable disease or adverse health effect may be related to the exposure;

(ii) there is a reasonable likelihood that the disease or effect may occur under the particular conditions of his work; and

(iii) there are valid techniques for detecting indications of the disease or the effect

and the technique of investigation is of low risk to the employee.

The remaining parts of *Regulation 11* (*Paragraphs (3)* to *(11)*) relate to the manner in which health surveillance is conducted and used, together with the maintenance of and access to surveillance records.

The decision as to whether health surveillance is appropriate to protect the health of employees is one that would normally be taken at the time of a COSHH assessment or during its subsequent review. However, for those processes and substances specified in *Schedule 6* to the Regulations surveillance must be carried out – in the main continuing requirements which were in place prior to the *COSHH Regulations*.

Normally health surveillance programmes would be initiated and carried out under the overall supervision of a registered medical practitioner, and preferably one with relevant occupational health experience. However, the surveillance itself (as described below) may be carried out by an occupational health nurse, a technician or a responsible member of staff, providing that individual was competent for the purpose.

There are many different procedures available for health surveillance including the following.

Biological monitoring

Measurement of the concentrations of hazardous substances or their metabolites within the body through testing of:

- blood, e.g. for lead or solvents;
- urine, e.g. for fluoride or solvents;
- exhaled air, e.g. for carbon monoxide, other gases or solvent vapours.

Biological effect monitoring

The measurement of the effects of hazardous substances on exposed workers, e.g. possible deterioration in the lungs through lung function and/or peak flow testing.

Medical surveillance

Physical examinations or measurements to identify possible alterations in body functions.

Interviews and/or examinations

Enquiries about possible symptoms by a suitably qualified person such as an occupational health nurse, e.g. the appearance of warts or lumps, possibly indicating skin cancer amongst pitch workers; inspections for possible ulceration amongst chrome workers.

For health surveillance to be 'appropriate' there must first be a significant enough risk to justify it and there must also be a valid technique (such as those above) for detecting indications of related occupational diseases or ill-health effects. The HSE provide specialist guidance on the subject (**REFS 37** and **38**).

Records of health surveillance must contain information set out in *Paragraphs 232 and 233* of the COSHH ACOP (**REF. 1**) and must be retained for at least 40 years.

Information, instruction and training

9.36 *Regulation 12* of the *COSHH Regulations* contains requirements relating to information, instruction and training for persons who may be exposed to substances hazardous to health.

(1) Every employer who undertakes work which is liable to expose an employee to a substances hazardous to health shall provide that employee with suitable and sufficient information, instruction and training.

(2) Without prejudice to the generality of paragraph (1), the information, instruction and training provided under that paragraph shall include —

(a) details of the substances hazardous to health to which the employee is liable to be exposed including —

(i) the names of those substances and the risk which they present to health,

(ii) any relevant workplace exposure limit or similar occupational exposure limit,

(iii) access to any relevant safety data sheet, and

(iv) other legislative provisions which concern the hazardous properties of those substances;

(b) the significant findings of the risk assessment;

(c) the appropriate precautions and actions to be taken by the employee in order to safeguard himself and other employees at the workplace;

(d) the results of any monitoring of exposure in accordance with regulation 10 and, in particular, in the case of any substance hazardous to health for which a workplace exposure limit has been approved, the employee or his representatives shall be informed forthwith, if the results of such monitoring show that the workplace exposure limit is exceeded;

(e) the collective results of any health surveillance undertaken in accordance with regulation 11 in a form calculated to prevent those results from being identified as relating to a particular person; and

(f) where employees are working with a Group 4 biological agent or material that may contain such an agent, the provision of written instructions and, if appropriate, the display of notices which outline the procedures for handling such an agent or material.

(3) The information, instruction and training required by paragraph (1) shall be—

(a) adapted to take account of significant changes in the type of work carried out or methods of work used by the employer; and

(b) provided in a manner appropriate to the level, type and duration of exposure identified by the risk assessment.

(4) Every employer shall ensure that any person (whether or not his employee) who carries out work in connection with the employer's duties under these Regulations has suitable and sufficient information, instruction and training.

(5) Where containers and pipes for substances hazardous to health used at work are not marked in accordance with any relevant legislation listed in Schedule 7, the employer shall, without prejudice to any derogations provided for in that legislation, ensure that the contents of those containers and pipes, together with the nature of those contents and any associated hazards, are clearly identifiable.

The general requirements of *Paragraph (1)* match those found in various other regulations but the contents of *Paragraph (2)* are much more prescriptive than the requirements of previous versions of the *COSHH Regulations*. Whilst it is important that workers are aware of the substances they are exposed to and the risks that they present, only a minority are likely to comprehend fully the significance of WELs, or understand all of the contents of a typical safety data sheet. A good awareness of the risks will mean employees are more likely to take the appropriate precautions. Instruction and training will be particularly relevant in relation to:

- awareness of safe systems of work;
- correct use of LEV equipment;
- use, adjustment and maintenance of PPE (especially RPE);
- the importance of good personal hygiene standards;
- requirements relating to eating, drinking and smoking;
- emergency procedures;
- arrangements for cleaning and the disposal of waste;
- contents of containers and pipes (see *Paragraph (5)* of *Regulation 12* above).

Employers must also inform employees or their representatives of the results of workplace exposure monitoring – forthwith if the WEL is shown to have been exceeded. Employees must also be informed of the *collective* results of health surveillance, e.g. the average urinary fluoride concentration within a department or on a particular shift, or the numbers of employees referred for further investigation following a skin inspection.

The type of information, instruction and training provided must be appropriate for the level, type and exposure involved and adapted to take account of significant changes.

Any person carrying out work on the employer's behalf (whether or not an employee) is required by *Paragraph (4)* to have the necessary information, instruction and training. Thus a consultant making a COSHH assessment or an occupational hygienist conducting exposure monitoring must be verified by the employer as being competent for the purpose and be furnished with the information necessary to carry out their work effectively.

Accidents, incidents and emergencies

9.37 *Regulation 13* of the *COSHH Regulations* requires employers to ensure that arrangements are in place to deal with accidents, incidents and emergencies related to hazardous substances. (This is in addition to the general duty to have procedures 'in the event of serious and imminent danger' which is contained in *Regulation 8* of the *Management Regulations* – see **CHAPTER 2: WHAT THE MANAGEMENT REGULATIONS REQUIRE**.) Based on HSE guidance in the ACOP booklet (**REF. 1**) such emergencies might include:

- a serious process fire posing a serious risk to health;
- a serious spillage or flood of a corrosive agent;
- a failure to contain biological, carcinogenic or mutagenic agents;
- a failure that could lead (or has led) to a sudden release of chemicals;
- a threatened significant exposure over an OEL, e.g. due to a failure of LEV or other controls.

The Regulation is rather lengthy and prescriptive in its requirements which are summarised below.

Arrangements should include:

- emergency procedures, such as:
  - appropriate first-aid facilities;
  - relevant safety drills (tested at regular intervals);
- providing information on emergency arrangements:
  - including details of work hazards and emergencies likely to arise;
  - made available to relevant accident and emergency services;
  - displayed at the workplace, if appropriate;

- suitable warning and other communications:
  - to enable an appropriate response, including remedial actions and rescue operations.

  In the event of an accident, incident or emergency, the employer must:

- take immediate steps to:
  - mitigate its effects;
  - restore the situation to normal;
  - inform employees who may be affected;
- ensure only essential persons are permitted in the affected area and that they are provided with:
  - appropriate PPE;
  - any necessary safety equipment and plant;
- in the case of a serious biological incident, inform employees or their representatives, as soon as practicable of:
  - the causes of the incident or accident;
  - the measures taken or being taken to rectify the situation.

Paragraph (4) of *Regulation 13* states that such emergency arrangements are not required where the risk assessment shows that because of the quantities of hazardous substances the risks are slight and control measures are sufficient to control that risk. (However, this 'exception' does not apply in the case of carcinogens or biological agents.)

Paragraph (5) of *Regulation 13* requires employees to report possible releases of a biological agent which could cause severe human disease forthwith to their employer (or another employee with specific responsibility).

The HSE guidance in the ACOP booklet (**REF. 1**) states that whether or not arrangements are required under *Regulation 13* is a matter for judgement, based on the potential size and severity of accidents and emergencies which may occur. Many incidents will be capable of being dealt with by the control measures required by *Regulation* 7. Even if an emergency does occur, the response should be proportionate – a small leak would not necessarily justify an evacuation of the workplace. A major section of the ACOP booklet provides further details of what emergency procedures might need to include, such as:

- the identity, location and quantities of hazardous substances present;
- foreseeable types of accidents, incidents or emergencies;
- special arrangements for emergencies not covered by general procedures;
- emergency equipment and PPE, and who is authorised to use it;
- first-aid facilities;

- emergency management responsibilities, e.g. emergency controllers;
- how employees should respond to incidents;
- clear up and disposal arrangements;
- arrangements for regular drills or practice;
- dealing with special needs of disabled employees.

Review of assessments

9.38 *Paragraph (3)* of *Regulation 6* of the *COSHH Regulations* requires COSHH assessments to *be reviewed regularly and forthwith if*:

- there is reason to suspect the risk assessment is no longer valid;
- there has been a significant change in the work to which the risk assessment relates; or
- the results of any monitoring carried out in accordance with *Regulation 10* show it to be necessary.

Assessments no longer valid

9.39 Assessments might be shown to be no longer valid because of:

- new information received on health risks, e.g. information from suppliers, revised HSE guidance, changes in the WEL;
- results from inspections or thorough examinations or tests (*Regulation 9*), e.g. indicating fundamental flaws in engineering controls;
- reports or complaints about defects in control arrangements
- the results from workplace exposure monitoring (*Regulation 10*), e.g. showing the WEL is regularly being exceeded (or approached);
- results from health surveillance (*Regulation 11*), e.g. demonstrating an unsatisfactory or deteriorating position;
- a confirmed case of an occupational disease.

Significant changes

9.40 Changes necessitating a review of an assessment might involve:

- the substances used, their form or their source;
- equipment used for the process or activity (including control measures);
- methods of work or operational procedures;

- volume, rate or type of production;
- staffing levels and related practical difficulties or pressures.

Regular review

9.41 The periods elapsing between reviews should relate to the degree of risk involved and the nature of the work itself. The COSHH ACOP previously stated that assessments should be reviewed at least every five years, but no longer contains such a recommendation. However, this would seem to be a reasonable maximum period to use.

A review would not necessarily require a revision of the assessment – it may conclude that existing controls are still adequate despite changed circumstances. However, where changes are shown to be required, *Paragraph (3)* of *Regulation 6* of the *COSHH Regulations* requires that these be implemented.

Some pitfalls

9.42 The author has encountered many situations where employers have not met the objectives of the *COSHH Regulations*, sometimes despite the expenditure of considerable time and effort. Amongst the more common pitfalls are the following.

The data sheet library

9.43 Employers have collected a vast library of manufacturers and suppliers data sheets (many of which relate to substances they do not use) in the belief that this constitutes a COSHH assessment. Whilst acquiring relevant data sheets is an important preparatory step and their availability can assist in respect of emergency medical treatment, simply acquiring data sheets is a long way short of what is required.

Armchair assessments

9.44 Some employers have been known to simply arrange for information from data sheets to be transcribed onto their own COSHH assessment form without anyone actually visiting the workplace to review the circumstances of use. Observations and discussions in the workplace are an essential part of assessing what control measures are necessary for individual processes or activities.

Overkill

9.45 Manufacturers and suppliers often identify precautions which *may* be appropriate for the use of their products. The unquestioning adoption of these precautions can often result in precautions (particularly PPE requirements) which are unnecessary, impractical for employees and extremely expensive. Whilst advice in data sheets is undoubtedly relevant, it is ultimately for the assessor(s) to determine what is appropriate in their workplace for their working methods and the quantity of substances they use.

Not seeing the wood for the trees

9.46 Particularly in larger organisations where large numbers of hazardous substances are in use, assessors often try to assess each and every substance individually. Assessing processes or activities is much more productive – looking at the layout and density of the forest rather than individual trees. Where a variety of substances are used in a process or activity, adopting a 'worst first' approach can often pay off. If control measures are adequate for the most hazardous substances they are likely to be adequate for other similar substances used in the same way.

Slaves to record systems

9.47 There is no single correct method for recording COSHH assessments – providing the appropriate information is included, many different formats can be used (as the examples in this chapter demonstrate). The imposition of standard forms (particularly over-complicated ones) can be a recipe for much grief and unnecessary work.

References

(All HSE publications)

9.48

1	L5	Control of substances hazardous to health. COSHH Regulations 2002. ACOP and Guidance (2005)
2	L8	Legionnaires' disease. The control of legionella bacteria in water systems.ACOP and Guidance (2000)
3	L9	Safe use of pesticides for non-agricultural purposes. ACOP (1995)
4	L55	Preventing asthma at work. How to control respiratory sensitisers (1994)
5	L59	Control of substances hazardous to health in the production of pottery. ACOP (1995)
6	L86	Control of substances hazardous to health in fumigation operations. ACOP (1996)
7	HSG 110	Seven steps to successful substitution of hazardous substances (1994)
8	HSG 125	A brief guide on COSHH for the offshore oil and gas industry (1995)

9	HSG 188	Health risk management. A guide to working with solvents (1999)
10	INDG 95	Respiratory sensitisers and COSHH (1995)
11	INDG 136	COSHH. A brief guide to the Regulations (2005)
12	INDG 198	Working with sewage: the health hazards. A guide for employers (1995)
13	INDG 233	Preventing dermatitis at work (1996)
14	INDG 257	Pesticides: Use them safely (1997)
15	INDG 273	Working safely with solvents (1998)
16	INDG 311	Beryllium and you (2000)
17	INDG 315	Stone dust and you (2001)
18	INDG 319	Working safely with coating powders (2000)
19	INDG 329	Benzene and you (2000)
20	INDG 346	Chromium and you (2001)
21	INDG 351	Nickel and you (2002)
22	INDG 353	Why do I need a safety data sheet? (2002)
23	INDG 391	Cadmium and you (2004)
24	MSA 8	Arsenic and you (1991)
25	MSA 17	Cobalt and you (1995)
26	MSA 19	PCBs and you (1995)
27	MSA 21	MbOCA and you (1996)
28	MSB 4	Skin cancer caused by pitch and tar (1996)
29	HSG 37	An introduction to local exhaust ventilation (1993)
30	HSG 53	Respiratory protective equipment at work (2005)
31	HSG 206	Cost and effectiveness of chemical protective gloves for the workplace (2001)
32	INDG 330	Selecting protective gloves for work with chemicals (2000)
33	L25	Personal protective equipment at work (2005)
34	EH 40	Workplace exposure limits (updated regularly, often annually)
35	HSG 193	COSHH Essentials. Easy steps to control chemicals (2003)
36	HSG 54	The maintenance, examination and testing of local exhaust ventilation (1998)
37	HSG 61	Health surveillance at work (1999)
38	HSG 167	Biological monitoring in the workplace (1997)

10 Noise assessment

In this chapter:

Introduction	**10.1**
How noise damages hearing	**10.2**
Damage to the hearing cells	**10.2**
Other damage	**10.3**
Noise measurement	**10.4**
The decibel scale	**10.4**
Personal noise exposure	**10.5**
Noise measurement instruments	**10.6**
The Control of Noise at Work Regulations 2005 summarised	**10.7**
Commencement, interpretation and application	**10.8**
Assessment of the risk created by exposure to noise	**10.9**
Elimination or control of exposure to noise	**10.10**
Hearing protection	**10.11**
Maintenance and use of equipment	**10.12**
Health surveillance	**10.13**
Information, instruction and training	**10.14**
Exemptions	**10.15**
Noise exposure reduction	**10.16**
Alternative work methods	**10.17**
Reducing noise generation	**10.18**
Reducing noise transmission	**10.19**
Equipment specification	**10.20**
Ear protection	**10.21**
Planning and preparation	**10.22**
Who will carry out the assessment?	**10.23**
How will the assessment be organised?	**10.24**
Gathering information	**10.25**
Making the assessment	**10.26**
The purpose of the assessment	**10.26**
Assessment methods and considerations	**10.27**
Noise assessment in the field	**10.28**
Drawing conclusions	**10.29**
Assessment records	**10.30**
After the assessment	**10.31**
Review and implementation of recommendations	**10.31**
Possible health surveillance	**10.32**
Assessment review	**10.33**
References	**10.34**

Introduction

10.1 The risks associated with exposure to noise have been understood for many years. HM Factory Inspectorate (a forerunner of the HSE) published its booklet 'Noise and the Worker' in 1963 and the courts have deemed that employers should have been aware of the risks since that date and been taking action to control them. However, employers were slow to move and as a result there have been many successful compensation claims for occupational deafness and many workers are also in receipt of government disablement benefit in respect of noise-induced hearing loss.

Occupational deafness has occurred in many industries – mining, shipbuilding, engineering, textiles, construction and the wood trades being some of the more common. However, there was no specific legislation dealing with noise until 1990 when the *Noise at Work Regulations 1989* came into operation. These Regulations were replaced by the *Control of Noise at Work Regulations 2005* which came into operation on 6 April 2006, although there were transitional provisions for some work sectors. At the heart of the Regulations is a requirement for employers to make an assessment of the risks from noise.

How noise damages hearing

Damage to the hearing cells

10.2 Hearing takes place through the action of sound waves on the eardrum. The eardrum vibrates, activating the bones of the inner ear, which in turn exert pressure on the cochlea, a snail-shaped organ containing a liquid. Motion of liquid within the cochlea is detected by tiny hair cells which transmit sound to the brain. It is these hair cells which are damaged by excessive exposure to noise.

An analogy is to consider the hair cells as similar to grass on a lawn. One person's path across the lawn may be visible for a short while but the grass will soon return to its previous position. The path of several people walking across the lawn will be visible for rather longer but will have disappeared by the following day. However, a large number of people walking to and fro on a daily basis will eventually damage the grass beyond its ability to recover.

Similarly the hair cells in the ear may suffer a short-term threshold shift from which they can recover but longer exposure to higher levels of noise will cause permanent damage to the hearing. Usually the frequencies at the higher end of the scale are affected first. This means that the victim can still hear sounds but loses the ability to differentiate between them, particularly the consonants. This may result in accusations that others are not speaking clearly rather than an acceptance that hearing damage has taken place. Many individuals also become quite adept at lip-reading to overcome their hearing loss. A certain amount of hearing loss also occurs naturally as part of the ageing process.

Other damage

10.3 Very loud noises (such as explosions and gunfire) can cause perforated ear drums although this is relatively rare and the eardrum usually heals up. Cases of occupational deafness are often accompanied by tinnitus – a continuous ringing, buzzing or whistling in the ear – although this may have other causes. Noise may also create other work-related problems not associated with hearing damage – it may hinder communication, cause stress and sudden loud noises may startle workers. Whilst outside the scope of the *Control of Noise at Work Regulations 2005*, significant risks of these types should still be assessed as part of the employer's general risk assessment.

Noise measurement

The decibel scale

10.4 The ear can hear sounds at frequencies between 20 and 20 000 cycles per second or Hertz (Hz). It is most sensitive to frequencies between 3000 and 6000 Hz, those used in human speech. Sound pressure levels are measured in decibels (dB) with the range going from zero decibels (the threshold of hearing) up to around 140 decibels (a very painful and dangerous level of exposure). A correction is made to allow for the human ear's varying ability to hear sounds at different frequencies. This is called the 'A weighting' and noise levels corrected in this way are shown in dB(A). The 2005 Regulations introduced the concept of the 'peak sound pressure level'. This is measured using the 'C weighting', shown as dB(C).

The decibel scale is logarithmic. Consequently a rise of 10 dB(A) (e.g. from 90 dB(A) to 100 dB(A)) actually represents a tenfold increase in noise. An increase of 3 dB(A) results approximately in a doubling of the noise. This feature of the decibel scale is particularly important in assessing the risks from noise – an apparently small increase in dB(A) can in fact significantly increase the risks.

Most noises in the workplace are a mixture of a wide range of frequencies – pure tone noises (i.e. noise of a single frequency) are usually only generated by test instruments or tuning forks. Instruments normally measure a combination of all of the noise frequencies they are exposed to. However, more sophisticated instruments (see later) can select noise from particular frequencies. This is known as octave band analysis.

An approximate guide to typical sound levels is provided in the table below:

dB(A)	
0	Faintest audible sounds
10	Leaf rustling, quiet whisper
20	Very quiet room
30	Subdued speech
40	Quiet office
50	Normal conversation
60	Busy office
70	Loud radio or TV
80	Busy street
90	Heavy vehicle close by
100	Road drill
110	Chainsaw
120	Riveting
140	Jet aircraft taking off close by

Personal noise exposure

10.5 *Regulation 4* of the *Control of Noise at Work Regulations 2005* sets out exposure limit values and action values on which other requirements of the Regulations are based.

The **lower exposure action values** are:

- a daily or weekly personal noise exposure level of 80 dB(A); and
- a peak sound pressure of 135 dB(C).

The **upper exposure action values** are:

- a daily or weekly personal noise exposure level of 85 dB(A); and
- a peak sound pressure of 137 dB(C).

The **exposure limit values** are:

- a daily or weekly personal noise exposure level of 87 dB(A); and
- a peak sound pressure of 140 dB(C).

Weekly exposure values may be used where exposure varies markedly from day to day. In applying the exposure limit values (which must not be exceeded) account can be taken of hearing protection. (The exposure action values have both been reduced by 5 dB(A) from their equivalents in the 1989 regulations.)

In calculating daily personal noise exposures it is important to appreciate the logarithmic nature of the decibel scale. Each of the following exposures will produce a daily personal noise exposure of 85 dB(A) – assuming that the remainder of the working day is relatively quiet.

dB(A)	
85	8 hours
88	4 hours
91	2 hours
94	1 hour
104	6 minutes
114	36 seconds

Noise measurement instruments

10.6 In order to estimate noise exposure, it will be necessary to measure both A-weighted sound pressure levels and the maximum C-weighted peak sound pressure level. This can be done with an integrating sound level meter meeting at least Class 2 of BS EN ISO 61672 -1: 2003 (or type 2 of BS EN 60804: 2001, the former standard).

Personal sound exposure meters (dosemeters) can be used to measure the noise exposure of individuals, and are particularly useful where workers are mobile or access for measurement is difficult. Dosemeters should meet the standards of BS EN 61252: 1997. Dual purpose instruments can operate as both a sound level meter and a dosemeter.

Dosemeters are prone to being sabotaged through deliberate exposure to excessive noise. Data recorders used in conjunction with measurement instruments can be very useful in identifying when high exposure to noise may have occurred. Calibrators should be used to check meters each day both before and after use, and meters and calibrators should be tested at least every two years.

The Control of Noise at Work Regulations 2005 summarised

10.7 The Regulations in full are contained in the HSE booklet *'Controlling noise at work'* (**REF. 1**). This also provides detailed guidance on their interpretation and many practical aspects of noise measurement and control. Shorter guidance for employers is also available in a free HSE leaflet (**REF. 2**).

Commencement, interpretation and application

10.8 The Regulations came into force on 6 April 2006 with the exception of the music and entertainment sectors where they apply from 6 April 2008 (the 1989 Regulations continuing to apply in the interim). There are also delays and restrictions in application of the Regulations to ships. Various terms, including 'the music and entertainment sectors', are defined in *Regulation 2*.

Regulation 3 states that the Regulations are to protect persons against risks arising from their exposure to noise at work. Consequently, account does not have to be taken under these Regulations of risks to members of the public (including customers), although employers and the self-employed still have their general duties to such persons under *Section 3* of *HSWA 1974*. However, employers still have duties to workers other than their own employees under the Regulations, although these duties are qualified by the phrase 'so far as is reasonably practicable'.

Assessment of the risk created by exposure to noise

10.9 Where work is liable to expose any employees to noise at or above a lower exposure action value (defined in *Regulation 4* and referred to in 10.5 above), *Regulation 5* requires a 'suitable and sufficient' risk assessment to be carried out. The Regulation specifies how the assessment should be conducted and various factors which must be taken into consideration. These are explained more fully in 10.27 below.

Elimination or control of exposure to noise

10.10 *Paragraph (1)* of *Regulation 6* states that 'The employer shall ensure that risk from the exposure of his employees to noise is either eliminated at source or, where this is not reasonably practicable, reduced to as low a level as is reasonably practicable.' *Paragraph (2)* requires that where an employee is likely to be exposed to noise at or above an upper exposure level, exposure must be reduced to as low a level as is reasonably practicable by appropriate organisational and technical measures, excluding the provision of hearing protection. Actions which must be taken by the employer are specified in *Paragraph (3)* and must include consideration of:

- other working methods which reduce exposure to noise;
- choice of equipment emitting the least possible noise (see also 10.20);
- the design and layout of workplaces, workstations and rest facilities;
- information and training (see also 10.14);
- reduction of noise by technical means (see also 10.18 and 10.19);
- appropriate maintenance programmes;
- limitation of the duration and intensity of exposure;
- appropriate work schedules with adequate rest periods.

Employers must ensure that employees are not exposed to noise above an exposure limit value or, if an exposure limit is exceeded, they must forthwith:

- reduce exposure to below the exposure limit value;
- investigate the reason for the limit being exceeded; and
- modify the organisational and technical measures to prevent it being exceeded again.

Hearing protection

10.11 *Regulation* 7 requires that where employees are likely to be exposed to noise at or above a lower exposure action level, their employers must make hearing protection available on request. If levels of noise exposure are likely to be above an upper exposure action level then

- hearing protection must be provided;
- relevant areas must be designated as Hearing Protection Zones (indicated by the appropriate signs);
- access to such areas must be restricted, where practicable and justified by the risk;
- employers must ensure, so far as is reasonably practicable, that employees do not enter such areas without hearing protection.

Hearing protection is covered in more detail in 10.21.

Maintenance and use of equipment

10.12 Under *Regulation 8* employers must ensure (so far as is practicable) that anything they provide under the Regulations is fully and properly used. (This requirement excludes hearing protection, which is covered by *Regulation* 7.) All equipment (including hearing protection) must be 'maintained in an efficient state, in efficient working order and in good repair'.

Employees must make full and proper use of hearing protection provided (where noise levels are at or above an upper exposure action level) and any other control measures provided under the regulations, e.g. by closing doors of acoustic enclosures. They must also report any defects in hearing protection or other control measures as soon as practicable.

Health surveillance

10.13 *Regulation 9* introduces a new requirement for health surveillance where the risk assessment indicates that there is a risk to the health of employees exposed to noise. The HSE guidance (**REF. 1**) states that surveillance should be provided to workers regularly exposed above the upper exposure action values or to individuals who are particularly sensitive to noise who are exposed to lesser levels. Health surveillance is covered in more detail in 10.32.

Information, instruction and training

10.14 Employers have a duty under *Regulation 10* to provide employees likely to be exposed at or above a lower exposure action value (and their representatives) with suitable and sufficient information, 'instruction and training'. This must include:

- the nature of risks from noise exposure;
- organisational and technical control measures;
- exposure limit values and action values;
- significant findings of the risk assessment;
- availability and provision of hearing protection;
- why and how to detect and report signs of hearing damage;
- any entitlement to health surveillance;
- safe working practices to minimise exposure to noise;
- collective results of any health surveillance.

The HSE pocket card 'Protect your hearing or lose it' (**REF. 3**) may be useful in providing information to employees. Information, instruction and training must be updated to take account of significant changes. Employers must also ensure that persons carrying out work in connection with their duties (e.g. noise measurement or risk assessment) have suitable and sufficient information, instruction and training.

Exemptions

10.15 The Regulations also make provision for possible exemption from the Regulations:

- from hearing protection, where likely to cause greater risk to health or safety (*Regulation 11*);
- for the emergency services (*Regulation 12*); and
- relating to the Ministry of Defence (*Regulation 13*).

Noise exposure reduction

10.16 Before carrying out a noise assessment, it is important to have an awareness of the different measures which can be adopted to reduce the exposure of employees to noise, including the different types of hearing protection available. Whilst noise control engineering is a specialist subject in its own right, an appreciation of the techniques available can be very useful in identifying measures which may be effective in given situations and also in evaluating the recommendations being made by the so-called specialists. The HSE have published an excellent booklet called 'Sound Solutions' which contains 60 case studies of noise control techniques which have been used successfully in a wide variety of industrial situations (**REF. 4**). Similar booklets are available for some specific industries (**REFS 5** and **6**). Much useful information is also provided in 'Controlling Noise at Work' (**REF. 1**).

Alternative work methods

10.17 *Regulation 6* requires that risk from exposure to noise is eliminated at source, where reasonably practicable. There may be scope to use less noisy work methods or equipment, e.g.

- use of hydraulic pile driving in place of impact methods;
- high pressure water jetting rather than abrasive cleaning;
- welding or bolting fabrications together instead of riveting.

Reducing noise generation

10.18 Noise is generated by vibrating sources – either a vibrating surface or vibration in a fluid. If the vibration can be eliminated or reduced then so will the noise. It may be important to consider other parts which vibrate in sympathy (secondary vibration) as well as the primary source of the vibration. Noise reduction measures might include the following.

Vibrating surfaces

- cushioning impacts (e.g. with plastic, rubber or nylon surfaces);
- replacing metal gears by nylon/polyurethane gears or by belts;
- using isolating or anti-vibration mountings;
- separating large vibrating surfaces from moving parts;
- stiffening structural parts or panels;
- placing machines on vibration absorbing pads;
- using damping materials on metal surfaces;
- using mesh in place of sheet metal;
- placing absorbent gaskets around doors and lids;
- replacing rigid pipework and flexible materials.

Vibration in gases

- choosing centrifugal rather than propeller fans;
- using large diameter, low speed fans;
- using large diameter, low pressure ductwork;

- streamlining ductwork to avoid turbulence (this will also improve its efficiency);
- using effective silencers to reduce turbulence at exhausts;
- using low-noise air nozzles or pneumatic ejectors (at the minimum pressure necessary).

Efficient maintenance

Good maintenance standards are also important in reducing noise generation, e.g. by

- replacing worn or badly fitting parts;
- securing loose parts;
- balancing rotating and other moving parts correctly;
- providing good lubrication.

Reducing noise transmission

10.19 Many of the measures described to reduce noise generation will also reduce noise transmission within work equipment. However, it is also important to reduce the transmission of noise through the air.

Acoustic enclosures

Noisy equipment can be placed in acoustic enclosures. Alternatively where the noise source is large or there are several noise sources, it may be better to place the workers inside an acoustic enclosure. This may be an acoustically protected control booth, in which case it should contain all the relevant controls and provide adequate visibility of the equipment or process being controlled. (These are commonly used in large printing plants, e.g. in the newspaper industry.) In some cases the enclosure may simply be a noise refuge, to which workers can go when their duties do not require them to be out on the noisy plant (such refuges are often used in power generation plants).

Important considerations in designing acoustic enclosures are:

- covering the surfaces of the enclosure in sound absorbent material;
- minimising the openings in the enclosure;
- installing absorbent gaskets around doors, windows, service inlets etc.;
- avoiding the enclosure being in contact with vibrating parts.

Applying sound absorbent materials

Transmission of noise can be reduced by the application of sound absorbent materials to prevent noise reflection from the walls and ceilings of rooms. Such materials will generally be most efficient if they are installed close to the source of the noise but benefits can also be obtained from installing them close to positions where people work. Hanging panels of sound absorbent material close to noise sources may also be effective. Portable sound absorbent screens can also be useful, especially for protecting maintenance employees working for limited periods in close proximity to noisy equipment.

Separation measures

The transmission of noise to workers can be reduced by situating them further away from noise sources or, if possible, by putting them in different rooms. Care should always be taken in the positioning of noisy exhaust or extraction systems to ensure that noise (together with any gas, dust or fumes) is directed away from working positions.

Equipment specification

10.20 Manufacturers are under a legal obligation to design and construct machinery so as to reduce risks from noise. They must also provide information about noise emissions. This should enable potential purchasers to compare the noise from different products and select accordingly. 'Controlling Noise at Work' (**REF. 1**) provides considerable detail on this subject. However, there is a typical 'uncertainty' in the declared value of up to 3 dB and some noise test methods may not be representative of true working conditions. Caution should be exerted in interpreting test results – noise measurements should be made under normal operating conditions and preferably before equipment has been formally accepted. Similarly, caution should be exercised over performance standards quoted for acoustic enclosures.

Ear protection

10.21 There are various types of hearing protection available, all of which must satisfy the relevant part of BS EN 352.

Ear muffs

Ear muffs (also known as ear defenders) consist of hard plastic cups which fit over the ears. They are sealed against the head by cushions containing foam or a viscous liquid or gel. The inside part of the cups is filled with soft plastic foam or similar sound-absorbent material. The cups are pressed against the head by pressure bands which normally pass over the top of the head although they may be positioned behind the head or under the chin. Some types of ear muffs are designed to be attached to safety helmets.

The use of ear muffs over spectacles will decrease their efficiency as also may the presence of long hair or jewellery. Both the pressure exerted by the head band and the condition of the muffs seals are likely to deteriorate with age or misuse.

Ear plugs

Ear plugs are intended to be fitted directly into the ear canal. Some types are reusable whilst others are disposable. Reusable plugs are usually of plastic or rubber materials whilst disposable plugs are made of foam plastic or a down material coated in a plastic membrane. Some reusable plugs are attached to cords to prevent their loss (an important consideration in food handling).

Hygiene is very important in relation to ear plugs as contamination or infections can easily be introduced into the ear. Reusable plugs should always be washed prior to further use. Plugs should not be used by anyone suffering from an ear infection or irritation. Some employees are unable to tolerate anything within their ear at all.

Some reusable plugs come in different sizes – some workers need different sizes for each ear. Some types of plugs are moulded individually to fit a person's ears. Simple instruction is usually necessary to show employees how to insert the plugs into their ears safely and efficiently. Ear plugs are also available fitted to a neckband. These can easily be slipped off and carried around the neck, although the pressure exerted by the neckband may deteriorate with time.

Special types of protection

There are increasing numbers of specialist types of hearing protection including level-dependent protectors (which protect against high noise levels while permitting communication in quieter conditions) and active noise-reduction protectors (which incorporate an electronic sound cancelling system). Some forms of protection have communication facilities allowing the wearer to receive direct messages or signals and also entertainment programmes.

The choice of hearing protection will depend upon a number of factors including those described below.

Level of protection

All suppliers of ear protection must supply information on the level of protection their products provide. This will not be identical for all noise frequencies – most types are more effective against the frequencies in the audible range. There are different ways of expressing the degree of protection (attenuation) – all of which are described in the HSE guidance booklet (**REF. 1**).

Complicating the situation further, research has shown that the attenuation figures provided by suppliers are not always achieved in practice. This may be due to poor fitting, deterioration of the ear protection, use with spectacles, long hair or jewellery. The HSE recommend that what is described as a 'real world factor' of 4 dB is also applied.

Work limitations/personal preference

Many people find ear muffs more comfortable for wearing in situations where they are exposed to noise for long periods. Semi-inserts are convenient for work which involves regularly going into and out of hearing protection areas. Plugs can be kept in the pocket for use in situations of occasional or unexpected exposure to noise. Some individuals find the presence of anything in their ear canal to be irritating.

Assuming that the degree of protection provided by the different types is similar (and this is often the case) it is generally best to provide employees with the type of protection that they prefer. If hearing protection is only worn by an employee for 90% of the time they are exposed to noise, then the maximum exposure reduction which will be achieved overall can be shown mathematically to be only 10 dBA. If hearing protection is only worn for half of the exposure period then noise exposure will only be reduced by 3 dBA.

Cost

If hearing protection is not worn then any cost saving in opting for cheaper types is a false economy. Although disposable plugs may seem to be a relatively cheap option, they will be used in large numbers in workplaces where there is regular or continuous exposure to noise. Reusable types of hearing protection will usually prove more economical in the long run in these environments. However, disposable plugs (kept readily available in pockets) may be the better option for occasional exposure situations.

Planning and preparation

10.22 The principles to be followed in noise assessments are similar to those for general risk assessments (covered in **CHAPTER 4: CARRYING OUT RISK ASSESSMENTS**) and for most of the other more specific type of assessments. Some specific factors relating to noise assessments are referred to below.

Who will carry out the assessment?

10.23 The assessment must be carried out by a competent person, and noise assessments probably require a greater degree of technical expertise than other types of risk assessment. The HSE guidance (**REF. 1**) refers to skills and knowledge needed as including:

- understanding how risks can arise from noise exposure;
- being able to identify potentially problematic noise sources;

- understanding noise information from machinery suppliers;
- understanding the work going on;
- being able to estimate noise exposure and make related judgements;
- understanding exposure action and limit values, and legal requirements;
- understanding good practice and industry standards for noise control;
- being able to prioritise controls and tackle immediate risks;
- recognising the need for and being able to access further specific skills or advice.

That is not to say that noise assessment is solely the preserve of specialists. Straightforward situations, particularly those involving steady noise, can be assessed by anyone with an understanding of the decibel scale who is capable of using simple instruments and relating sound levels to the requirements of the Regulations. However, more complex noise exposure will require a greater degree of knowledge and experience in its assessment. Creation of an in-house assessment team may be appropriate in larger workplaces.

How will the assessment be organised?

10.24 As with other types of assessment, some workplaces may need to be divided into manageable assessment units. These may be based on:

- Departments or sections.
- Buildings or rooms.
- Process lines.
- Use of types of equipment.
- Specific activities.

Some areas or activities may be ruled out of the assessment process because it is known that noise exposure is below the lower exposure action value. The HSE guidance booklet provides some simple tests to determine whether a risk assessment is required.

Test	Probable noise level	Assessment needed if noise like this for more than
Noise intrusive but normal conversation possible	80 dB	6 hours
You have to shout to talk to someone 2 m away	85 dB	2 hours
You have to shout to talk to someone 1 m away	90 dB	45 mins

At the other end of the scale it may be decided that the noise exposure is sufficiently complex to require competent outside assistance in order to make an adequate assessment. However, employers should be cautious in contracting out noise assessments. There have been many instances where outside consultancies have carried out extensive measurement of noise levels in workplaces using some very sophisticated instrumentation and at considerable expense, but without getting round to answering the basic questions that the assessment process should be addressing.

It will also be necessary at this stage to check that the noise measuring instruments likely to be required during the assessment are not only available but are also in full working order with up-to-date calibration certificates.

Gathering information

10.25 Prior to starting the assessment relevant information should be gathered such as:

- What equipment, processes, activities, areas are thought to be noisy.
- Previous noise surveys or previous noise assessments.
- Noise specifications for relevant equipment.
- HSE guidance material – (**REFS 5–16** relate to specific occupational sectors and types of equipment).

Making the assessment

The purpose of the assessment

10.26 In making a noise assessment it is important to keep in mind the basic purpose of the assessment required by *Regulation 4*. In essence, answers must be provided to the following questions:

- Is there a risk due to noise in this situation?
- Who is exposed and to what extent? (in particular, in relation to the exposure limit values and action values).
- What must be done to comply with the Regulations?
  - elimination or control of noise exposure (equipment specification, noise generation, noise transmission);
  - provision of hearing protection;
  - creation of hearing protection zones;
  - provision of information, instruction and training to employees.

Whilst noise survey work will often need to be carried out, this should only be to the extent required to answer the above questions. Extremely detailed noise surveys are not a requirement of the Regulations in their own right.

Assessment methods and considerations

10.27 *Regulation 5* states that assessment of the levels of noise to which workers are exposed to must be by means of:

- observation of specific working practices;
- reference to relevant information on the probable levels of noise (see 10.25 above);
- if necessary, measurement of noise levels.

It goes on to say that the assessment must include consideration of:

- the level, type and duration of exposure;
- the effects of exposure on employees or groups of employees at particular risk;
- any interaction between noise and ototoxic substances or vibration;
- any interaction between noise and audible warning signals etc.;
- information provided by equipment manufacturers;
- availability of alternative equipment to reduce noise emission;
- any exposure beyond normal working hours or in rest facilities;
- information from health surveillance, including published information;
- the availability of adequate hearing protection.

Noise assessment in the field

10.28 Of these considerations, the level, type and duration of exposure are probably the most important. By a combination of observations and discussions with workers and their representatives it will be necessary to determine:

- What equipment, processes, activities, areas are noisy?
- What are the actual noise levels?
- Do the noise levels vary?
  - at different stages of activities;
  - for different materials, products, equipment.

- How long is noise emitted at each level?
- How long are workers exposed to each level of noise?
- What is their position in relation to the noise source?
- Are different workers exposed to different levels of noise?

Each of these questions need not be answered precisely, but enough information must be obtained in order to decide whether the exposure reaches either the lower or the upper exposure action value or even if there is a risk of the exposure limit value being exceeded. The HSE booklet (**REF. 1**) contains much guidance on practical ways of determining exposures where noise levels vary.

During the assessment the availability and effectiveness of existing noise exposure control measures should also be evaluated, e.g.:

- acoustic enclosures;
- types of hearing protection provided;
- designated hearing protection zones and related signs;
- compliance with hearing protection requirements;
- quality of equipment maintenance.

Drawing conclusions

10.29 At the end of the assessment the assessors must make conclusions in respect of:

Where the problems are

- which equipment, processes, activities, areas involve exposure above the exposure action values.

Noise reduction measures

- existing measures which are proving effective;
- additional measures which should be introduced;
- alternative possibilities requiring evaluation.

Hearing protection

- existing hearing protection zones;
- the need for additional zones;

- types of hearing protection provided;
- enforcement of hearing protection requirements;
- the quality, wording and positioning of signs;
- the supply and maintenance of hearing protection.

Information, instruction and training

- whether workers need further information on noise risks (both general and specific);
- are there any specific noise-related training needs;
- whether workers need reminding of their obligations under the regulations.

Assessment records

10.30 *Regulation 5* of the *Control of Noise at Work Regulations 2005* requires that the significant findings of the risk assessment are recorded as soon as practicable after the assessment is made or changed. The measures taken (or intended to be taken) to meet the requirements of *Regulations 6, 7* and *10* must also be recorded.

Records should include details of the workplaces, areas or tasks assessed, the date of the assessment, who carried out the assessment and what the results of the assessment were.

There is no standard format for assessment records but they might contain:

- noise exposure tables (identified by equipment, area, activity or person);
- plans showing noise exposures at various locations;
- details of exposure times;
- details of activities carried out by peripatetic workers (including noise exposure levels and times);
- recommendations on actions required;
- reference to more detailed assessments which may be required;
- details of noise control measures and hearing protection arrangements;
- details of information, instruction and training.

An example of an assessment record is provided later in this chapter.

SPHINX ENGINEERING
NOISE ASSESSMENT REPORT

Dates of assessment: 28 March 2006 *Assessment by*: D Barr

Instruments used: Acme Sound Level Meter Model D2
Reliant Personal Dosemeter Type R2

RESULTS OF NOISE SURVEY

Location	Sound Level	Average Daily Exposure	Assessed Exposure Level dB(A)			Comments
	dB(A)	(hours)	< 80	80–85	85+	
MACHINE SHOP						
Bandsaw	93	2			✓	Operators do not generally move between machines.
Power press A	94	4			✓	
Power press B	95	2			✓	Machines are well separated.
Milling Machine	82	6		✓		Some operators wearing hearing protection.
Drilling Machines	77	4	✓			No signs in position.
Background	76	8	✓			
PLASTIC SHOP						
Cross cut saw	95	1			✓	Dosemeter readings:
Portable saw	97	1			✓	89.7 dBA and 90.3 dB(A) (employee 1);
Assembly bench						88.6 dB(A) (employee 2).
– saw operating	92	2			✓	Employee 2 wears hearing protection, employee 1 does not. No signs.
– assembly only	75		✓			

Recommendations

1 Noise reduction

1.1 Investigate means of reducing noise exposure at

- the bandsaw and power presses in the Machine Shop;
- the saws in the Plastic Shop (see separate proposals).

2 Ear protection

2.1 Continue to make hearing protection (ear muffs and re-usable ear plugs) available to all employees;

2.2 remind the Purchasing Department that all hearing protection must meet BS EN 352.

3 Ear protection zones

3.1 Make the following compulsory hearing protection zones

- work at the bandsaw and power presses in the Machine Shop;
- work in all parts of the Plastic Shop whenever a saw is operating.

3.2 Provide suitable signs to indicate hearing protection zones (as above) on the bandsaw and power presses (Machine Shop) and at the entrance to the Plastic Shop.

4 Information to employees

4.1 At the next staff meeting

- show the video about risks from noise;
- explain the results of the noise survey;
- emphasise the compulsory hearing protection zones;
- show the hearing protection available and how to use it.

After the assessment

Review and implementation of recommendations

10.31 CHAPTER 4: CARRYING OUT RISK ASSESSMENTS provides general guidance on the review of recommendations and the implementation of an action plan resulting from risk assessments. The nature of noise problems is such that specialist external resources may be required to provide guidance on the feasibility or practicalities of noise control measures or to conduct detailed noise surveys or assessments. This may result in the timescale for implementation of some actions being rather longer than for other types of risk assessment.

Recommendations should also be followed up in order to ensure that they have actually been implemented and, where relevant, to measure the new levels of noise exposure. Either the previous assessment records should be annotated with the findings of the follow-up or a revised assessment record should be prepared.

Possible health surveillance

10.32 Health surveillance is required by *Regulation 9(1)* 'If the risk assessment indicates that there is a risk to the health of . . . employees who are, or are liable to be, exposed to noise . . .'. Guidance in *Controlling Noise at Work* (**REF. 1**) states that such surveillance should be provided 'to workers regularly exposed above the upper exposure action values'. Those individuals who may be particularly sensitive to noise and are exposed between the lower and the upper exposure action values or occasionally above the upper exposure action value should also be subject to surveillance.

Surveillance must include the testing of hearing (audiometry), and attendance is compulsory for employees subject to this requirement. Records must be kept of surveillance, with employees being allowed access to their personal record on request. Analysis of anonymous results of groups of employees can be useful to the employer in monitoring the effectiveness of noise control and hearing conservation activities.

Where, as a result of surveillance, an employee is found to have identifiable hearing damage, the employer must ensure that the individual is examined by a doctor. If the doctor (or any subsequent specialist) considers the damage likely to be the result of exposure to noise, the employer must:

- ensure that a suitably qualified person informs the employee;
- review the risk assessment;
- review measures taken to comply with *Regulations 6, 7* and *8* (noise control, hearing protection and their maintenance and use);
- consider redeploying the employee to alternative work;
- ensure continued surveillance;
- review the health of other similarly exposed employees.

Audiometric testing must only be carried out by persons who have had appropriate training. Where employers do not have competent occupational health staff within their own organisation, they are likely to need the service of an external specialist consultancy. Detailed guidance on audiometric testing and related records is provided in *Controlling Noise at Work* (**REF. 1**).

Assessment review

10.33 *Regulation 5(4)* requires the assessment to be reviewed regularly, and forthwith if there is reason to suspect it is no longer valid or significant changes have taken place. Circumstances justifying a review might include:

- evidence of hearing loss (e.g. from audiometric tests on employees);
- changes in work equipment or equipment layout;

- changes in workload, work pattern or machine speeds;
- changes in materials used or products manufactured;
- significant alterations in employees' duties or hours of work.

Regular reviews may detect the cumulative effects of minor changes or of wear and tear in equipment. Such reviews may involve simple spot checks rather than detailed reviews or re-assessments.

References

(All HSE publications)

10.34

1	L108	Controlling Noise at Work (2005)
2	INDG 362	Noise at work. Guidance for employers on the Control of Noise at Work Regulations 2005 (2005) – free leaflet.
3	INDG 363	Protect your hearing or lose it (2005) – free pocket card.
4	HSG 138	Sound solutions. Techniques to reduce noise at work (1995)
5	HSG 232	Sound solutions for the food and drink industries. Reducing noise in food and drink manufacturing (2002)
6	HSG 182	Sound solutions offshore. Practical examples of noise reduction (1998)
7	INDG 127	Noise in construction. Further guidance on the Noise at Work Regulations 1989 (1995)
8	HSG 109	Control of noise in quarries (1993)
9	PM 56	Noise from pneumatic systems (1985)
10	PBIS 1	Noise assessments in paper mills (2000) – free leaflet.
11	AS 8	Noise (in agriculture) (2002) – free leaflet.
12	WIS 4	Noise reduction at band re-saws (1990) – free leaflet.
13	WIS 5	Noise enclosure at band re-saws (1992) – free leaflet.
14	WIS 13	Noise at woodworking machines (1997) – free leaflet.
15	EIS 26	Noise in engineering (1998)
16	FIS 32	Reducing noise exposure in the food and drink industries (2002)

11 Assessment of manual handling

In this chapter:

Introduction 11.1

What the Regulations require 11.2

Risk of injury from manual handling 11.3
Assessment guidelines 11.3
Task-related factors 11.4
Load-related factors 11.5
Factors relating to the working environment 11.6
Capability of the individual 11.7

Avoiding or reducing risks 11.8
Elimination of handling 11.9
Automation or mechanisation 11.10
Reduction of the load 11.11
Task-related measures 11.12
Load-related measures 11.13
Improving the work environment 11.14
Individual capability 11.15

Planning and preparation 11.16
Who will carry out the assessments? 11.17
How will the assessments be arranged? 11.18
Gathering information 11.19

Making the assessment 11.20
Observations 11.21
Discussions 11.22
Notes 11.23

After the assessment 11.24
Assessment records 11.24
Review and implementation of recommendations 11.25
Assessment review 11.26
Training 11.27

References 11.28

Introduction

11.1 The *Manual Handling Operations Regulations 1992* came into operation on 1 January 1993 as part of the so-called 'six pack' of new Regulations. Although the Regulations implemented *European Directive 90/269/EEC*, the

HSE had previously attempted to introduce Regulations on manual handling in an attempt to reduce the huge toll of accidents from this source. At the time of the introduction of the Regulations more than a quarter of the accidents reported each year to the enforcing authorities (the HSE and local authorities) were associated with manual handling. In 1990/91, 65% of handling accidents resulted in sprains and strains, whilst 5% were fractures. Many manual handling injuries have cumulative effects eventually resulting in physical impairment or even permanent disability.

There is already evidence of the beneficial effects of the Regulations, particularly in respect of the increasing use of mechanical handling aids and improved manual handling training. By 1996/97 the proportion of handling accidents causing fractures had been reduced to 3.3%. However, the battle was far from over with 54% of RIDDOR reportable accidents in the health and social work sector still attributable to handling. Several other sectors had over 30% of accidents involving handling, including food and drink, textiles, construction, wholesale and retail, transport/storage and education.

What the Regulations require

11.2 Various terms used in the Regulations are defined in *Regulation 2* which states:

> 'manual handling operations' means any transporting or supporting of a load (including the lifting, putting down, pushing, pulling, carrying or moving thereof) by hand or by bodily force.
>
> 'Injury' does not include injury caused by any toxic or corrosive substance which:
>
> (a) has leaked or spilled from a load;
>
> (b) is present on the surface of a load but has not leaked or spilled from it; or
>
> (c) is a constituent part of a load;
>
> and 'injured' shall be construed accordingly;

(This in effect establishes a demarcation with the *COSHH Regulations* – where there is oil on the surface of a load making it difficult to handle that is a matter for the *Manual Handling Operations Regulations*, whereas any risk of dermatitis is covered by the *COSHH Regulations*).

> Load includes any person and any animal

(This is of great importance to a number of work sectors including health and social care, agriculture and veterinary work).

The HSE guidance accompanying the Regulations states that an implement, tool or machinery being used for its intended purpose is not considered to constitute a load – presumably establishing a demarcation with the requirements of PUWER 1998. However, when such equipment is being moved before or after use it will undoubtedly be a load for the purpose of these Regulations.

Regulation 2(2) imposes the duties of an employer to his employees under the Regulations on a self-employed person in respect of himself. *Regulation 3* excludes the normal ship-board activities of a ship's crew under the direction of the master from the application of the Regulations (these are subject to separate Merchant Shipping legislation).

The duties of employers under the Regulations are all contained in *Regulation 4* which (as in the *COSHH Regulations*) establishes a hierarchy of measures that employers must take. *Regulation 4* states,

(1) Each employer shall –

(a) so far as is reasonably practicable, avoid the need for his employees to undertake any manual handling operations at work which involve a risk of their being injured; or

(b) where it is not reasonably practicable to avoid the need for his employees to undertake any manual handling operations at work which involve a risk of their being injured–

(i) make a suitable and sufficient assessment of all such manual handling operations to be undertaken by them, having regard to the factors which are specified in column 1 of Schedule 1 to these Regulations and considering the questions which are specified in the corresponding entry in column 2 of that Schedule,

(ii) take appropriate steps to reduce the risk of injury to those employees arising out of their undertaking any such manual handling operations to the lowest level reasonably practicable, and

(iii) take appropriate steps to provide any of those employees who are undertaking any such manual handling operations with general indications and, where it is reasonably practicable to do so, precise information on –

(aa) the weight of each load, and

(bb) the heaviest side of any load whose centre of gravity is not positioned centrally.

(2) Any assessment such as is referred to in paragraph (1)(b)(i) of this Regulation shall be reviewed by the employer who made it if –

(a) there is reason to suspect that it is no longer valid; or

(b) there has been a significant change in the manual handling operations to which it relates,

and where as a result of any such review changes to an assessment are required, the relevant employer shall make them.

As a result of the *Health and Safety (Miscellaneous Amendments) Regulations 2002*, a further paragraph was added to *Regulation 4*. This states:

(3) In determining for the purpose of this regulation whether manual handling operations at work involve a risk of injury and in determining the appropriate steps to reduce that risk regard shall be had in particular to—

(a) the physical suitability of the employee to carry out the operations;

(b) the clothing, footwear or other personal effects he is wearing;

(c) his knowledge and training;

(d) the results of any relevant risk assessment carried out pursuant to regulation 3 of the Management of Health and Safety at Work Regulations 1999;

(e) whether the employee is within a group of employees identified by that assessment as being especially at risk; and

(f) the results of any health surveillance provided pursuant to regulation 6 of the Management of Health and Safety Regulations 1999.

The hierarchy of measures which must be taken by employers where there is a risk of injury can be summarised as:

- avoid the operation (if reasonably practicable);
- assess the remaining operations;
- reduce the risk (to the lowest level reasonably practicable);
- inform employees about weights.

Reference will be made later in the chapter to a variety of measures which can be taken to avoid or reduce risks. The factors to which regard must be given during the assessment (as specified in *Schedule 1* to the Regulations) are the *tasks, loads, working environment* and *individual capability* together with other factors referred to in *Regulation 4(3)*. These factors and the questions specified by *Schedule 1* are also covered later in the chapter.

Regulation 5 places a duty on employees. It states:

> Each employee while at work shall make full and proper use of any system of work provided for his use by his employer in compliance with *Regulation 4(1)(b)(ii)* of the *Regulations*.

Regulation 6 provides for the Secretary of State for Defence to make exemptions from the requirements of the Regulations in respect of home forces and visiting forces and their headquarters.

Regulation 7 extends the Regulations to apply to offshore activities such as oil and gas installations, including diving and other support vessels. Previous provisions relating to manual handling were replaced or revoked by *Regulation 8*. The Regulations themselves are extremely brief but they are accompanied by considerable HSE guidance, much of which will be referred to during this chapter.

Risk of injury from manual handling

Assessment guidelines

11.3 Appendix 1 of the guidance accompanying the Regulations (**REF. 1**) provides assessment guidelines which can be used to filter out manual handling operations involving little or no risk and helps to identify where a detailed risk assessment is necessary. However, it must be stressed that *these are not weight limits* – they must not be regarded as safe weights nor must they be treated as thresholds which must not be exceeded. The HSE guidance states that application of the guidelines will provide a reasonable level of protection to around 95% of working men and women. Some workpeople will require additional protection (e.g. pregnant women, frail or elderly workers) and this should be considered in relation to the assessment of individual capability.

Lifting and lowering

The guidelines for lifting and lowering operations are shown in the accompanying diagram. They assume that the load can be easily grasped with both hands and that the operation takes place in reasonable working conditions, using a stable body position. Where the hands enter more than one box, the lowest weights apply. A detailed assessment should be made if the weight guidelines are exceeded or if the hands move beyond the box zones.

These guideline weights apply up to approximately 30 operations per hour. They should be reduced for more frequent operations:

- by 30% for 1 or 2 operations per minute;
- by 50% for 5 to 8 operations per minute;
- by 80% for more than about 12 operations per minute.

Other factors likely to require a more detailed assessment are where:

- workers do not control the pace of work, e.g. on assembly lines;
- pauses for rest are inadequate;
- there is no change of activity allowing different muscles to be used;
- the load must be supported for any length of time.

Carrying

Similar guidelines figures can be applied for carrying where the load is held against the body and carried up to 10 metres without resting. For longer distances or loads held below knuckle height, a more detailed assessment should be made. The guideline figures can be applied for carrying loads on the shoulder for distances in excess of 10 metres but an assessment may be required for the lifting of the load onto and off the shoulder.

Pushing and pulling

Where loads are slid, rolled or supported on wheels the guideline figures assume that force is applied with the hands between knuckle and shoulder height. The guideline figure for starting or stopping the load is a force of about 25 kg (about 250 Newtons) and 10 kg (about 100 Newtons) for keeping the load in motion for men (these figures reduce to 16 and 7 kg respectively for women).

In practice it is extremely difficult to make a judgement as to whether forces of these magnitude are being applied. Many employers are likely to make assessments of significant pushing or pulling activities anyway. There are no guideline limits on distances over which loads may be pushed or pulled, providing there is sufficient opportunity for rest.

Handling when seated

The accompanying diagram illustrates the guideline figures for handling whilst seated. Any handling activities outside the specified box zones should be assessed in any case.

Schedule 1 to the Regulations contains a list of questions which must be considered during assessments of manual handling operations. All of these indicate factors which may increase the risk of injury. (Most of these factors have been included in the checklist for assessments provided later in the chapter.)

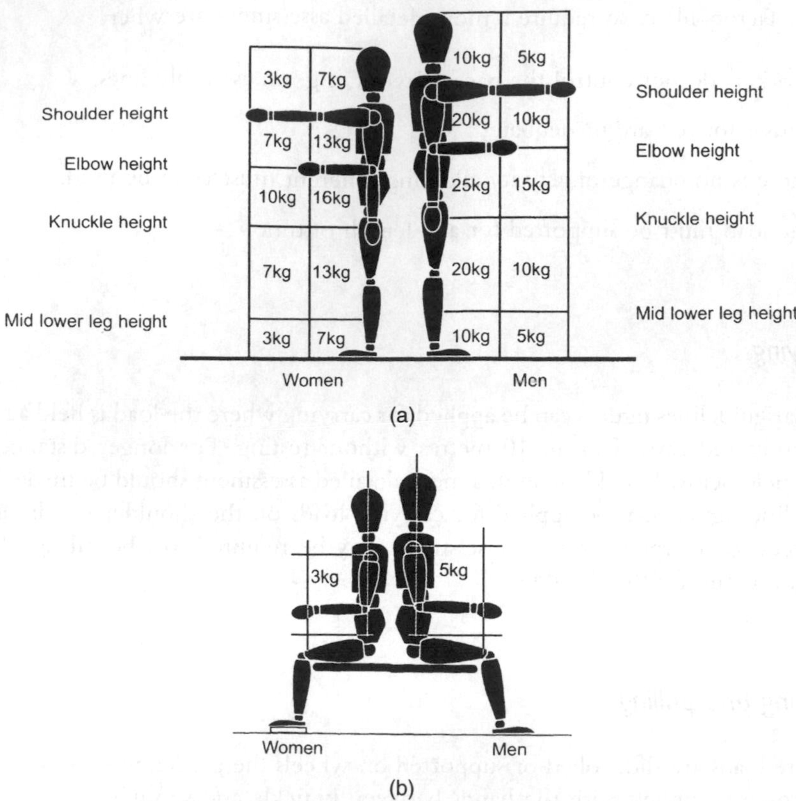

Figure 11.3

Task-related factors

Holding or manipulating loads at a distance from the trunk

11.4 Stress on the lower back increases as the load is moved away from the trunk. Holding a load at arm's length imposes five times the stress as the same load very close to the trunk.

Unsatisfactory body movement or posture

The risk of injury increases with poor feet or hand placings, e.g. feet too close together, body weight forward on the toes, heels off the ground.

Twisting the trunk whilst supporting a load increases stress on the lower back – the principle should be to move the feet, not twist the body.

Stooping also increases the stress on the lower back whilst *reaching upwards* places more stress on the arms and back and lessens control on the load.

Excessive movement of loads

Risks increase the further loads must be *lifted or lowered*, especially when lifting from floor level. Movements involving a *change of grip* are particularly risky. Excessive *carrying* of loads increases fatigue – hence the guidance figure of about 10 metres carrying distance before a more detailed assessment is made.

Excessive pushing or pulling

Here the risks relate to the forces which must be exerted, particularly when starting to move the load, and the quality of the grip of the handler's feet on the floor. Risks also increase when pushing or pulling below knuckle height or above shoulder height.

Risk of sudden movement

Freeing a box jammed on a shelf or a jammed machine component can impose unpredictable stresses on the body, especially if the handler's posture is unstable. Unexpected movement from a client in the health care sector or an animal in agricultural work can produce similar effects.

Frequent or prolonged physical effort

Risk of injury increases when the body is allowed to become tired, particularly where the work involves constant repetition or a relatively fixed posture. The periodic inter-changing of tasks within a work group can do much to reduce such risks.

Clothing, footwear, etc.

The effects of any PPE or other types of clothing, footwear, etc. required for the task may make manual handling more difficult or increase the possibilities of fatigue, dehydration, etc.

Insufficient rest and recovery

The opportunity for rest and recovery will also help to reduce risks, especially in relation to tasks which are physically demanding.

Rate of work imposed by a process

In such situations workers are often not able to take even short breaks, or to stretch or exercise different muscle groups. More detailed assessments should be made of assembly line or production line activities. Again periodic role changes within a work team may provide a solution.

Load-related factors

Is the load too heavy?

11.5 The guideline figures provided earlier are relevant here but only form part of the picture. Task-related factors (e.g. twisting or repetition) may also come into play or the load may be handled by two or more people. Size, shape or rigidity of the load may also be important.

Is the load bulky or unwieldy?

Large loads will hinder close approach for lifting, be more difficult to get a good grip of or restrict vision whilst moving. The HSE guidance suggests that any dimension exceeding 75 cm will increase risks and the risks will be even greater if it is exceeded in more than one dimension. The positioning of suitable handholds on the load may somewhat reduce the risks.

Other factors to consider are the effects of wind on a large load, the possibility of the load hitting obstructions or loads with offset centres of gravity (not immediately apparent if they are in sealed packages).

Is the load difficult to grasp?

Where loads are large, rounded, smooth, wet or greasy, inefficient grip positions are likely to be necessary, requiring additional strength of grip. Apart from the possibility of the grip slipping there are likely to be inadvertent changes of grip posture, both of which could result in loss of control of the load.

Is the load unstable?

The handling of people or animals was referred to earlier. Not only do such loads lack rigidity, but they may also be unpredictable and protection of the load as well as the handler is of course an important consideration. Other potentially unstable loads are those where the load itself or its packaging may disintegrate under its own weight or where the contents may shift suddenly, e.g. a stack of books inside a partially full box.

Is the load sharp, hot, etc.?

Loads may have sharp edges or rough surfaces or be extremely hot or cold. Direct injury may be prevented by the use of protective gloves or clothing but the possibility of indirect risks (e.g. where sharp edges encourage an unsuitable grip position or cold objects are held away from the body) must not be overlooked.

Factors relating to the working environment

Space constraints

11.6 Areas of restricted headroom (often the case in maintenance work) or low work surfaces will force stooping during manual handling and obstructions (e.g. in front of shelves) will result in other unsatisfactory postures. Restricted working areas or the lack of clear gangways will increase risks in moving loads, especially heavier or bulkier items.

Uneven, slippery or unstable floors

All of the above can increase the risks of slips, trips and falls. The availability of a firm footing is a major factor in good handling technique – such a footing will be a rarity on a muddy construction site. Moving workplaces (e.g. trains, boats, elevating work platforms) will also introduce unpredictability in footing. (The guideline weights should be reduced in such situations.)

Variation in levels of floors or work surfaces

Risks will increase where loads have to be handled on slopes, steps or ladders, particularly if these are steep. Any slipperiness of their surface, e.g. due to ice, rain, mud will create further risks. The need to maintain a firm handhold on ladders or steep stairways is another factor to consider. Movement of loads between surfaces or shelving may also need to be considered, especially if there is considerable height change, e.g. floor level to above shoulder height.

Extremes of temperature or humidity

High temperatures or humidity will increase the risk of fatigue, with perspiration possibly affecting the grip. Work at low temperatures (e.g. in cold stores or cold weather) is likely to result in reduced flexibility and dexterity. The need for gloves and bad weather clothing is another factor to consider.

Ventilation problems or gusts of wind

Inadequate ventilation may increase fatigue whilst strong gusts of wind may cause considerable danger with larger loads, e.g. panels being moved on building sites.

Poor lighting

Lack of adequate lighting may cause poor posture or an increased risk of tripping. It can also prevent workers identifying risks associated with individual loads, e.g. sharp edges or corners, offset centres of gravity.

Capability of the individual

11.7 Manual handling capabilities will vary significantly between individual workers. Reference was made earlier in the chapter to the assessment guideline weights providing a reasonable level of protection to around 95% of workers and those guideline figures differed for men and women. *Paragraph 3* of *Regulation 4* now specifically refers to factors which must be taken into account in respect of individual workers. Individual factors include the following.

Gender

Whilst there is overlap between the capabilities of men and women, generally the lifting strength of women is less than men.

Age

The bodies of younger workers will not have matured to full strength (see **CHAPTER 3: SPECIAL CASES**) whilst with older workers there will be a gradual decline in their strength and stamina, particularly from the mid-1940s onwards. Such persons will be the types of groups of employees referred to in *Subsection (e)* of *Regulation 4(3)*.

Experience

More mature workers may be better able to recognise their own capabilities and pace themselves accordingly, whilst also having acquired better handling techniques.

Pregnancy

The need to protect new or expectant mothers was emphasised in **CHAPTER 3: SPECIAL CASES**.

Previous injury or ill-health

Workers with long-term musculoskeletal problems, a history of injuries or ill-health or short-term injuries (including those from non-work related causes) will all justify additional protection. Such problems should be detected by the health surveillance required by *Regulation 6* of the *Management Regulations* (see **11.15** later in the chapter).

Physique and stature

These factors will vary considerably within any workforce. Whilst bigger will often mean stronger, it must not be overlooked that some tasks or work locations may force taller workers to stoop or adopt other unsuitable handling postures.

Generally the objective should be to ensure that all manual handling operations can be performed satisfactorily by most reasonably fit and healthy employees, although a minority of workers may justify restrictions on their handling activities (see later in the chapter).

During the assessment process it is important that special note is taken of any manual handling operations which:

- Require unusual strength, height, etc., e.g. reels of wrapping paper can only be safely positioned in a machine by persons above a certain height.
- Create additional risk to workers who are pregnant, disabled or have health problems.

  Such tasks may need to be reassessed to reduce risks or it may be necessary to restrict some individuals from carrying them out.
- Require special information or training.

  The health care sector provides many examples of situations where both the worker and the client (the load) may be at risk if correct technique is not used. Many other tasks are likely to require specific techniques to be adopted – whether in manual handling itself or the use of handling aids.

Clothing, footwear, etc.

Some individuals may be affected more than others by the use of PPE or clothing, e.g. gloves, protective suits, breathing apparatus. These may significantly affect the user's mobility or dexterity in respect of manual handling. Similarly uniforms or costumes (e.g. in the entertainment industry) may also be a factor to be taken into account.

Avoiding or reducing risks

11.8 The Regulations require employers to avoid the need for employees to undertake manual handling operations involving a risk of injury, so far as is reasonably practicable. Measures for avoiding such risks can be categorised as:

- Elimination of handling.
- Automation or mechanisation.
- Reduction of the load.

Elimination of handling

11.9 It may be possible for employers to eliminate manual handling altogether (or to eliminate handling by their own employees) by means such as the following.

Redesigning processes or activities

For example, so that activities such as machining or wrapping are carried out in situ, rather than a product being manually handled to a position where the activity take place; a treatment is taken to a patient rather than vice versa (this may have other benefits to the patient).

Using transport better

For example, allowing maintenance staff to drive vehicles carrying tools or equipment up to where they are working, rather than manually handling them into place.

Requiring direct deliveries

For example, requiring suppliers to make deliveries directly into a store rather than leaving items in a reception area (such suppliers may be better equipped in terms of handling aids and their staff may be better trained in respect of manual handling).

Automation or mechanisation

11.10 The automation or mechanisation of a work activity is best considered at the design stage although there is no reason in principle why such changes cannot be made later, e.g. as the result of a risk assessment. In making changes it is important to avoid creating additional risks, e.g. the use of fork lift trucks for a task in an already cramped workplace may not be the best solution. However, as employers have become more aware of manual handling risks (many prompted by the advent of the Regulations) a much greater range of mechanical handling solutions to manual handling problems have become available.

These include the following.

Mechanical lifting devices

Fork lift trucks, mobile cranes, lorry-mounted cranes, vacuum devices and other powered handling equipment are commonly used to eliminate or greatly reduce the manual handling of loads.

Manually operated lifting devices

Pallet and stacker trucks, manually operated chain blocks or lever hoists can all be used to move heavy loads using very limited physical force.

Powered conveyors

As well as fixed conveyors (both belt and driven roller types), mobile conveyors are being used increasingly, particularly for the loading and unloading of vehicles, e.g. stacking of bundles of newspapers in delivery vans.

Non-powered conveyors, chutes, etc.

Free-running roller conveyors, chutes or floor-mounted trolleys allow loads to move under the effects of gravity or to be moved manually with little effort. Such devices may operate between different levels or be set into the workplace floor.

Trolleys and trucks

Trolleys and trucks can be used to greatly reduce the manual handling effort required in transporting loads. Some types also incorporate manual or mechanically powered lifting mechanisms. Specialist trolleys are designed for carrying drums or other containers whilst an ingenious triple wheel system can be lifted to some trucks to aid them in climbing or descending stairs.

Lifting tools

Special lifting tools are available to reduce the manual handling effort required for lifting or lowering certain loads, e.g. paving slabs, manhole covers, drums or logs.

All of these and many other types of devices are illustrated in an HSE booklet 'Manual handling: solutions you can handle' (**REF. 2**).

Reduction of the load

11.11 There is considerable scope to avoid the risk of injury by reducing the sizes of the loads which have to be handled. In some cases the initiative for this has come from suppliers of a particular product, whereas in others the impetus has come from customers or from users of equipment, often as a result of their manual handling assessments. Examples include:

- Packaged building materials (e.g. cement) reduced to 25 kg.
- Photocopying paper now supplied in 5 ream boxes (approximately 12 kg) – 10 ream boxes were previously commonplace.

- Newspaper bundle sizes reduced below 20 kg (printers reduce the number of copies in a bundle as the number of pages in the paper increases).
- Customers specifying maximum container weights they will accept from suppliers, e.g. weights of brochures from commercial printers.
- Equipment being separated into component parts and assembled where it is to be used, e.g. emergency equipment used, by fire services.

Where it is not reasonably practicable to avoid a risk of injury from an manual handling operation, the employer must carry out an assessment of such operations with the aim of reducing the risks to the lowest level reasonably practicable. The four main factors to be taken into account during the assessment are *the task, the load, the working environment* and *individual capability*. Possible risk reduction measures related to each of these factors are described below.

(Later in the chapter these are summarised in an assessment checklist.)

Task-related measures

Mechanical assistance

11.12 The types of mechanical assistance described earlier can be used to reduce the risk of injury, even if such risks cannot be avoided entirely. The use of roller conveyors, trolleys, trucks or levers can reduce considerably the amount of force required in manual handling tasks.

Task layout and design

The layout of tasks, storage areas, etc. should be designed so that the body can be used in its more efficient modes, e.g.:

- heavier items stored and moved between shoulder and mid-lower leg height;
- allowing loads to be held close to the body;
- avoiding reaching movements, e.g. over obstacles or into deep bins;
- avoiding twisting movements or handling in stooped positions, e.g. by repositioning work surfaces, storage areas or machinery;
- providing resting places to aid grip changes whilst handling;
- reducing lifting and carrying of loads, e.g. by pushing, pulling, sliding or rolling techniques;
- utilising the powerful leg muscles rather than arms or shoulders;
- avoiding the need for sustaining fixed postures, e.g. in holding or supporting a load;
- avoiding the need for handling in seated positions.

Work routines

Work routines may need to be altered to reduce the risk of injury using such measures as:

- limiting the frequency of handling loads (especially those that are heavier or bulkier);
- ensuring that workers are able to take rest breaks (either formal breaks or informal rest periods as and when needed);
- introducing job rotation within work teams (allowing muscle groups to recover whilst other muscles are used or lighter tasks undertaken).

Team handling

Tasks which might be unsafe for one person might be successfully carried out by two or more. The HSE guidance indicates that the capability of a two-person team is approximately two-thirds of the total of their individual capabilities; and of a three person team approximately one half of their total. Other factors to take into account in team handling are:

- the availability of suitable handholds for all the team;
- whether steps or slopes have to be negotiated;
- possibilities of the team impeding each other's vision or movement;
- availability of sufficient space for all to operate effectively;
- the relative sizes and capabilities of team members.

Load-related measures

Weight or size reduction

11.13 The types of measures described earlier in the chapter might be adopted to reduce the risk of injury even if it is not possible to avoid the risk entirely. However, reducing the individual load size will increase the number of movements necessary to handle a large total load. This could result in a different type of fatigue and also the possibility of corners being cut to save time. Sizes of loads may also be reduced to make them easier to hold or bring them closer to the handler's body.

Making the load easier to grasp

Handles, grips or indents can all make loads easier to handle (as demonstrated by the handholds provided in office record storage boxes). It will often be easier handling a load placed in a container with good handholds than handling it alone without any secure grip points. The positioning of handles or handholds on a load can also influence the handling technique used, e.g. help avoid stooping. Handholds should be wide enough to accommodate the palm and deep enough to accommodate the knuckles (including gloves where these are worn).

The use of slings, carrying harnesses or bags can all assist in gaining a secure grip on loads and carrying them in a more comfortable and efficient position.

Making the load more stable

Loads which lack rigidity themselves or are in insecure packaging materials may need to be stabilised for handling purposes. Use of carrying slings, supporting boards or trays may be appropriate. Partially full containers may need to have their contents wedged into position to prevent them moving during handling.

Reducing other risks

Loads may need to be cleaned of dirt, oil, water or corrosive materials for safe handling to take place. Sharp corners or edges or rough surfaces may have to be removed or covered over. There may also be risks from very hot or very cold loads. In all cases the use of containers (possibly insulated), other handling aids or suitable gloves or other PPE may need to be considered.

Improving the work environment

Providing clear handling space

11.14 The provision of adequate gangways and sufficient space for handling activities to take place will be linked with general workplace safety issues. Low headroom, narrow doorways and congestion caused by equipment and stored materials are all to be avoided. Good housekeeping standards are essential to safe manual handling.

Floor condition and design

Even in temporary workplaces such as construction sites, every effort should be made to ensure that firm, even floors are provided where manual handling is to take place. Special measures may be necessary to allow water to drain away or to clear promptly any potentially slippery materials (e.g. food scraps in kitchens). Where risks are greater special slip-resistant surfaces may need to be considered. In outdoor workplaces the routine application of salt or sand to surfaces made slippery by ice or snow may need to be introduced.

Differing work levels

Measures may be necessary to reduce the risks caused by different work levels. Slopes may need to be made more gradual, steps may need to be provided or existing steps or stairways made wider to accommodate handling activities. Work benches may need to be modified to provide a uniform and convenient height.

Thermal environment and ventilation

Unsatisfactory temperatures, high humidity or poor ventilation may need to be overcome by improved environmental control measures or transferring work to a more suitable area. For work close to hot processes or equipment, in very cold conditions (e.g. refrigerated storage areas) or carried on out of doors, the use of suitable PPE may be necessary.

Gusts of wind

Where gusts of wind (or powerful ventilation systems) could affect larger loads, extra precautions, in addition to those normally taken, may be necessary, e.g.:

- using handling aids (e.g. trolleys);
- utilising team-handling techniques;
- adopting a different work position;
- following an alternative transportation route.

Lighting

Suitable lighting must be provided to permit handlers to see the load and the layout of the workplace and to make accurate judgements about distance, position and the condition of the load.

Individual capability

11.15 In ensuring that individuals are capable of carrying out manual handling activities in the course of their work, account must be taken both of factors relating to the individual and of those relating to the activities they are to perform. Steps which may be appropriate include:

Medical screening

Workers may be screened both prior to being offered employment and periodically thereafter to check whether they are physically capable of carrying out the full range of manual handling tasks in the workplace. In some extreme situations (e.g. the emergency services) an individual may be deemed unsuitable for employment if they are not sufficiently fit. However, legislation prevents employers discriminating against those with disabilities *without good reason*. In most workplaces the results of such screening may simply result in restrictions on the range of tasks individuals are allowed to perform.

Long-term restrictions

Long-term restrictions may be appropriate where workers are identified through screening programmes as not being capable of carrying out certain manual handling activities without undue risk, or they become incapable due to some long-term injury or the effects of the ageing process. It may be necessary to place long-term restrictions on young workers until they attain an appropriate age at which their capabilities can be re-assessed.

Short-term restrictions

Relatively short-term restrictions on manual handling activities may be necessary because of:

- pregnancy (or a particular stage of pregnancy) – see **CHAPTER 3: SPECIAL CASES**;
- recently having given birth (also referred to in **CHAPTER 3: SPECIAL CASES**);
- injury or illness of short duration.

Fitness programmes

An increasing number of employers actively encourage employers to be physically fit, thus reducing their risk of injury during manual handling activities. A number of Fire Services have particularly pro-active programmes.

General manual handling training

Many employers offer general training in manual handling, either to all employees or to all those employees expected to engage in significant manual handling work. Aspects to cover in such training include:

- recognition of potentially hazardous operations;
- avoiding manual handling hazards;
- dealing with unfamiliar handling operations;
- correct use of handling aids;
- proper use of PPE;
- working environment factors;

- importance of good housekeeping;
- factors affecting individual capability (knowing one's limitations);
- good handling technique (summarised in **REF. 3**).

Task-specific training and instruction

Training and instruction will often need to relate to the safety of specific manual handling tasks

- how to use specific handling aids;
- specific PPE requirements;
- specific handling techniques.

The health care sector contains many such specific training needs (**REFS 4** and **5**) but there are also likely to be many specific needs in other occupational areas. *Subsection (c)* of *Regulation 4(3)* now requires employees' level of knowledge and training to be considered during the risk assessment.

Planning and preparation

11.16 The principles of planning and preparation for manual handling assessments are no different from those for general risk assessments. Reference should be made to the relevant section of **CHAPTER 4: CARRYING OUT RISK ASSESSMENTS** if further guidance is required in addition to that provided below.

Who will carry out the assessments?

11.17 The HSE guidance contained in booklet L 23 (**REF. 1**) states that while one individual may be able to carry out an assessment in relatively straightforward cases, in others it may be more appropriate to establish an assessment team. It also refers to employers and managers with a practical understanding of the manual handling tasks to be performed, the loads to be handled, and the working environment being better able to conduct assessments than someone outside the organisation. (Although it is not stated in the guidance, in-house personnel will also usually be more aware of the capabilities of employees carrying out manual handling tasks.)

The guidance refers to the individual or team performing the assessments needing to possess the following knowledge and expertise:

- the requirements of the Regulations;
- the nature of the handling operations;

- a basic understanding of human capabilities;
- awareness of high risk activities;
- practical means of reducing risk.

The HSE also acknowledge that there will be situations where external assistance may be required. These might include:

- training of in-house assessors;
- assessing risks which are unusual or difficult to assess;
- re-designing equipment or layouts to reduce risks;
- training of staff in handling techniques.

How will the assessments be arranged?

11.18 As for general risk assessments, the workplace or work activities will often need to be divided into manageable assessment units. Depending on how work is organised, these might be based on:

- Departments or sections.
- Buildings or rooms.
- Parts of processes.
- Product lines.
- Work stations.
- Services provided.
- Job titles.

Members of the assessment team can then be allocated to the assessment units where they are best able to make a contribution.

Gathering information

11.19 It is important to gather information prior to the assessment in order to identify potentially high risk manual handling activities and the precautions which ought to be in place to reduce the risks. Relevant sources might include the following.

Accident investigation reports

Dependent upon the requirements of the organisation's accident investigation procedure, these may only include the more serious accidents resulting from manual handling operations.

Ill health records

These may reveal short absences or other incidents due to manual handling which may have escaped the accident reporting system. Enquiries may need to be made in respect of the health surveillance records of individuals (see **11.15** above).

Accident treatment records

Accident books or other treatment records may identify regularly recurring minor accidents which do not necessarily result in any absences from work, e.g. cuts or abrasions from sharp or rough loads.

Operating procedures, etc.

Reference to manual handling activities and precautions which should be adopted may be contained in operating procedures, safety handbooks, etc.

Work sector information

Trade associations and similar bodies publish information on manual handling risks and how to overcome them.

HSE Guidance

Much guidance is available from the HSE – a number of publications are identified in the reference section of this chapter. Some material (**REFS 1–4**) will be relevant to most work activities, whilst others (**REFS 5–19**) relate to specific work sectors. The leaflet *Manual handling assessment charts* (**REF. 4**) demonstrates how quantitative methods can be used to assess manual handling risks, although the author does not particularly recommend this approach.

Making the assessment

11.20 As with other types of assessments, visits to the workplace are an essential part of manual handling assessment. Manual handling operations need to be observed (sometimes in some detail) and discussions with those carrying out manual handling work, their safety representatives and their supervisors or managers need to take place.

The assessment checklist provided on the following pages provides guidance on the factors which may need to be taken into account during the observations and discussions and also on possible measures to reduce the risk. The checklist utilises the four main factors specified in the Regulations – task, load, working environment and individual capability. In practice each of these will have greater or lesser importance, depending upon the manual handling activity.

MANUAL HANDLING ASSESSMENT CHECKLIST

THE TASK

ASSESSMENT FACTORS	***REDUCING THE RISK***
DISTANCE OF THE LOAD FROM THE TRUNK	TASK LAYOUT
BODY MOVEMENT/POSTURE – Twisting – Reaching – Stooping – Sitting	USE THE BODY MORE EFFECTIVELY e.g. SLIDING OR ROLLING THE LOAD SPECIAL SEATS
DISTANCE OF MOVEMENT	RESTING PLACE/TECHNIQUE
– Height (? Grip Change)	USE OF TROLLEYS, etc.
– Carrying (? Over 10m) – Pushing Or Pulling	TECHNIQUE/FLOOR SURFACE
RISK OF SUDDEN MOVEMENT	AWARENESS/TRAINING
FREQUENT/PROLONGED EFFORT	IMPROVED WORK ROUTINE e.g. Job rotation
CLOTHING/FOOTWEAR ISSUES	
RATE OF WORK IMPOSED BY A PROCESS	ADEQUATE REST OR RECOVERY PERIODS

THE LOAD

ASSESSMENT FACTORS	***REDUCING THE RISK***
HEAVY	MAKE THE LOAD LIGHTER TEAM LIFTING MECHANICAL AIDS
BULKY (Any Dimension above 75 cm)	MAKE THE LOAD SMALLER
UNWIELDY (Offset Centre Of Gravity)	INFORMATION/MARKING
DIFFICULT TO GRASP (Large, Rounded, Smooth, Wet, Greasy)	MAKE IT EASIER TO GRASP – Handles – Handgrips – Indents – Slings – Carrying Devices – Clean The Load
UNSTABLE – Contents Liable to Shift – Lacking Rigidity	WELL-FILLED CONTAINERS USE OF PACKING MATERIALS SLINGS/CARRYING AIDS
SHARP EDGES/ROUGH SURFACES	AVOID OR REDUCE THEM GLOVES OR OTHER PPE
HOT OR COLD	CONTAINERS (? Insulated)

WORKING ENVIRONMENT	
ASSESSMENT FACTORS	***REDUCING THE RISK***
SPACE CONSTRAINTS AFFECTING POSTURE	ADEQUATE GANGWAYS, FLOORSPACE, HEADROOM
UNEVEN, SLIPPERY OR UNSTABLE FLOORS	WELL-MAINTAINED SURFACES PROVISION OF DRAINAGE SLIP-RESISTANT SURFACING GOOD HOUSEKEEPING PROMPT SPILLAGE CLEARANCE
VARIATION IN WORK SURFACE LEVEL e.g. steps, slopes, benches	STEPS NOT TOO STEEP GENTLE SLOPES UNIFORM BENCH HEIGHT
HIGH OR LOW TEMPERATURES HIGH HUMIDITY POOR VENTILATION	BETTER ENVIRONMENTAL CONTROL RELOCATING THE WORK SUITABLE CLOTHING
STRONG WINDS OR OTHER AIR MOVEMENT	RELOCATING THE WORK ALTERNATIVE ROUTE HANDLING AIDS TEAM HANDLING
LIGHTING	SUFFICIENT WELL-DIRECTED LIGHT
MOVING WORKPLACE – Boat – Train – Vehicle	

INDIVIDUAL CAPABILITY	
ASSESSMENT FACTORS	***REDUCING THE RISK***
INDIVIDUAL FACTORS STRENGTH HEIGHT FLEXIBILITY STAMINA PROBLEMS WITH CLOTHING, ETC.	BALANCED WORK TEAMS 'SELF-SELECTION' MAKE ASSISTANCE AVAILABLE ENCOURAGE FITNESS
PREGNANCY INJURY/HEALTH PROBLEM – Back Trouble – Hernia – Temporary Injury	SPECIAL ASSESSMENTS FORMAL RESTRICTIONS
TASK FACTORS OPERATIONS OR SITUATIONS REQUIRING PARTICULAR – Awareness of risks – Knowledge – Method of approach – Technique	FORMALISED WORK PROCEDURES CLEAR INSTRUCTIONS ADEQUATE TRAINING – Recognising potentially hazardous operations – Dealing with unfamiliar operations – Correct use of handling aids – Proper use of PPE – Environmental factors – Importance of housekeeping – Knowing one's own limitations – Good technique

Observations

11.21 Sufficient time should be spent in the workplace to observe a sufficient range of the manual handling activities taking place and to take account of possible variations in the loads handled (raw materials, finished product, equipment, etc). Aspects to particularly be considered include:

- comparison of actual methods used with those specified in operating procedures, industry standards, etc.;
- the level and manner of use of handling aids and their effectiveness;
- handling techniques adopted;
- physical conditions, e.g. access, housekeeping, floor surfaces, lighting;
- variations between employees, e.g. size, strength, technique adopted.

Discussions

11.22 In discussing manual handling issues with workpeople a similar approach should be taken to that described in **CHAPTER 4: CARRYING OUT RISK ASSESSMENTS**, i.e. utilising open-ended questions wherever possible. Aspects which might be discussed include:

- whether normal conditions are being observed;
- their awareness of procedures, rules, etc.;
- their training in the use of handling aids or handling techniques;
- possible variations in work activities, e.g. product changes, seasonal variations, differences between day and night shift;
- what happens if handling equipment or handling aids break down or are unavailable;
- whether assistance is available if required;
- what manual handling problems have been experienced?;
- have there been any injuries with manual handling?;
- reasons precautions are not being taken.

Notes

11.23 Notes should be taken during the assessment of anything considered to be of relevance. Further guidance is provided in the equivalent section of **CHAPTER 4: CARRYING OUT RISK ASSESSMENTS** on what may be of relevance and on note taking techniques.

After the assessment

Assessment records

11.24 The HSE guidance states that the significant findings of the assessment must be recorded unless:

- the assessment could very easily be repeated and explained at any time because it is so simple and obvious; or
- the manual handling operations are quite straightforward, of low risk, are going to last only a very short time, and the time taken to record them would be disproportionate.

As has been recommended in relation to other types of assessment, it is better to make a simple record if there is any doubt.

There is no standard format for recording an assessment. The HSE Guidance booklet L 23 (**REF. 1**) includes a sample record format together with a useful worked example. This record format includes a checklist of assessment factors under the four main headings (task, load, working environment, capability) in a similar listing to the assessment checklist we have provided. Alongside each checklist item the assessor has space to:

- identify whether the risk from this factor is low, medium or high;
- provide a more detailed description of the problems;
- identify possible remedial action.

Some will no doubt find this checklist format meets their needs. However, the author has found that many assessment factors often have little or no relevance to a specific manual handling operation. As a result, the eventual record contains a lot of blank paper, with only occasional comments. Consequently he has preferred to keep the checklist separate and only include within the record a reference to the risks, precautions and recommended improvements which are relevant to the operation being assessed. Two worked examples of this type of record are included in the chapter.

MANUAL HANDLING ASSESSMENT		AREA: *Stores*	DATE: *24/3/00*	ASSESSOR: *B Strong*
No.	OPERATION	RISKS	EXISTING PRECAUTIONS	RECOMMENDATIONS
1	HANDLING OF INCOMING PARTS AND MATERIALS	Some items are both large and heavy, e.g. drums and some boxed materials. Palletised items must be transferred onto trolley by hand. (no fork-lift access into stores)	Vehicles are unloaded as close as possible to stores entrance. Trolley and sack barrow available and used. Heavy items are lifted by two people. (Stores staff have been trained in handling techniques).	Provide a suitable drum handling trolley. Investigate which items can be delivered in smaller containers. Provide a hand-operated pallet truck.
2	MOVEMENT OF PARTS AND MATERIALS WITHIN STORES	Transfer of items between stores sections (step between old and new sections causes problems). Movement of items onto and off shelves, etc.	Trolley and sack barrow and two-person lifts as above. Stores lighting is good. Heaviest items are mainly stored on shelves between knee and shoulder height. Clear access to shelves is maintained. Some items are rolled or pushed into fixtures.	Provide a gradual ramp alongside the step (or alternatively a 'climbing barrow'). Ensure assistance is available when required for moving heavier items. Some parts of the floor in the old store require repair.
3	DESPATCH OF ITEMS TO OPERATING DEPARTMENTS	A few heavy items are despatched, sometimes over long distances. Rough ground must be negotiated en route to annex building. Some transfers take place during hours of darkness.	Stores trolley and barrow are loaned, where necessary. Sometimes stores staff assist in moving loads.	Ensure all operating dept. staff collecting heavy items receive suitable training in handling techniques. Provide a suitable footpath to the annex. Check lighting levels of external areas.

MANUAL HANDLING ASSESSMENT		AREA: Newspaper Despatch†	DATE: 11/5/00	ASSESSOR: *A. Loader*
No.	OPERATION	RISKS	EXISTING PRECAUTIONS	RECOMMENDATIONS
	GENERAL COMMENTS	Weight of bundles controlled at 16–17 kg each.	Lighting and floor surface in despatch area and loading bay good.	
		Waste paper and strapping can create tripping risks.	Bins readily available for disposal of waste paper and strapping.	Enforce safe disposal of strapping more rigorously.
1	PREPARATION OF BUNDLES FOR DESPATCH	Manual removal and refeeding of mis-stacked and unstrapped bundles.	Automated line with conveyors between stacking, strapping and loading positions.	Provide a table close to the conveyor line for temporary storage.
		Some bundles are stored temporarily on the floor.	Staff trained in handling techniques.	
2	DELIVERY OF BUNDLES INTO VANS	Vans at bay C are loaded manually by company and contract delivery staff.	Adjustable mobile conveyors feed into vans at bays A and B.	
			Company staff trained in handling techniques.	Provide a similar mobile conveyor at bay C (see below also)
3	STACKING BUNDLES INSIDE VANS	Stacking may involve:	Company staff trained (see above).	Make company manual handling training available to contract staff.*
		stooping (due to low headroom)	Newer company vans now have adequate headroom.	Complete conversion of fleet by 2002.
		moving bundles at floor level	Bundles should be slid across the floor (rakes available).	Recommend contractors use similar vans.*
		moving bundles above shoulder height	This should normally be avoided by fully utilising the floor area.	

Notes: † A separate assessment of delivery activities would be necessary.
* The company has no legal obligation to do this.

Review and implementation of recommendations

11.25 The guidance provided in **CHAPTER 4: CARRYING OUT RISK ASSESSMENTS** in relation to the review of recommendations and implementation of action plans is equally applicable to manual handling assessments. Similarly recommendations will need to be followed up in order to ensure that they have actually been implemented and also to assess any new risks which may have been introduced. This latter point is particularly relevant to manual handling as changes in working practice may have resulted in different techniques being adopted (with their own attendant risks) or the introduction of handling aids may have created additional training needs.

Assessment review

11.26 *Paragraph 4(2)* of the Regulations requires a manual handling assessment to be reviewed if:

(a) there is reason to suspect that it is no longer valid; or

(b) there has been a significant change in the manual handling operations to which it relates;

As is the case for other assessments, a review would not necessarily result in the assessment being revised, although a revision may be deemed to be necessary. A significant manual handling–related injury should automatically result in a review of the relevant assessment. It is recommended that assessments are reviewed periodically in any case in order to take account of small changes in work practices which will eventually have a cumulative effect. The periods for such reviews might vary between two and five years – depending upon the degree of risk involved and the potential for gradual changes to take place.

Training

11.27 The need for specific manual handling technique training may frequently be one of the recommendations made as a result of a manual handling assessment. With manual handling activities featuring in a wide range of work sectors many employers have chosen to provide all or a significant proportion of their workforce with general training in good handling technique. Such training is often provided as part of a more comprehensive health and safety induction programme. Guidance on the content of such general training is contained in *Paragraphs 173* and *174* of the HSE Guidance booklet (**REF. 1**), *Getting to grips with manual handling* (**REF. 3**) and many other HSE publications.

References

(All HSE publications)

11.28

1	L 23	Manual Handling Operations 1992. Guidance on the Regulations (2004)
2	HSG 115	Manual Handling: Solutions you can handle (1994)
3	INDG 143	Getting to grips with manual handling: A short guide for employers (2005) – free leaflet.
4	INDG 383	Manual handling assessment charts (2003)
5		Manual handling in the health services (1998)
6		Getting to grips with handling problems: Worked examples of assessment and reduction of risk in the health service (1993)
7	HSG 119:	Manual handling in drinks deliveries (1994)
8	HSG 196	Moving food and drink: Manual handling solutions for the food and drink industries (2000)
9	HSG 171	Well handled: Offshore manual handling solutions (1997)
10		Picking up the pieces: Prevention of musculoskeletal disorders in the ceramics industry (1996)
11	AS 23	Manual handling solutions for farms (2000)
12	INDG 125	Handling and stacking bales in agriculture (1998) – free leaflet.
13	IACL 105:	Handling the news: Advice for employers on manual handling of bundles (1999) – free leaflet.
14	IACL 106	Handling the news: Advice for newsagents and employees on safe handling of bundles (1999) – free leaflet.
15	IACL 103	Manual handling in the textiles industry (1998)
16		Reducing manual handling injuries in the rubber industry (1999)
17	INDG 318	Manual handling solutions in woodworking (2000)
18	INDG 332	Manual packing in the brick industry (2000) – free leaflet
19	INDG 326	Manual handling in the railway industry. Advice for employers (2000) – free leaflet.

12 Assessment of DSE workstations

In this chapter:

Introduction	**12.1**
The Regulations summarised	**12.2**
Definitions	**12.2**
Exclusions	**12.3**
Assessment of workstations and reduction of risk	**12.4**
Requirements for workstations	**12.5**
Daily work routine of users	**12.6**
Eyes and eyesight	**12.7**
Provision of training	**12.8**
Provision of information	**12.9**
Planning and preparation	**12.10**
Who will carry out the assessments?	**12.10**
How will the assessments be organised?	**12.11**
Making the assessments	**12.12**
Observations	**12.12**
Discussions	**12.13**
Assessment records	**12.14**
Special situations	**12.15**
Homeworkers and teleworkers	**12.16**
Shared workstations	**12.17**
Work with portable DSE	**12.18**
Pointing devices	**12.19**
After the assessment	**12.20**
Review and implementation of recommendations	**12.20**
Assessment review/re-assessments	**12.21**
References	**12.22**

Introduction

12.1 The *Health and Safety (Display Screen Equipment) Regulations 1992*, together with the other 'six pack' *Regulations*, came into operation on 1 January 1993. Their aim was to combat upper limb pains and discomfort, eye and eyesight effects, together with general fatigue and stress associated with work at DSE. The HSE state in the guidance booklet to the Regulations (**REF. 1**) that they do not consider there are any radiation risks from DSE or special problems for pregnant women. The Regulations require Assessments to be carried out to identify any health and safety risks at workstations used by 'users' or 'operators' (as defined). 'Users' have the right to an eye and eyesight test and any spectacles (or contact lenses) found to be necessary for their DSE work must be provided by their employer.

Several minor amendments were made to the Regulations by the *Health and Safety (Miscellaneous Amendments) Regulations 2002*. The latest HSE guidance booklet (**REF. 1**) incorporates the amendments and also provides much more up-to-date guidance on many aspects of DSE work including homeworking and work with portable DSE such as laptop computers and portable phones with small screens.

The Regulations summarised

Definitions

12.2 *Regulation 1* contains a number of important definitions:

- Display screen equipment

  Any alphanumeric or graphic display screen, regardless of the display process involved.

- Workstation

  An assembly comprising DSE, any optional accessories, any disk drive, telephone, modem, printer, document holder, work chair, work desk, work surface or other peripheral item and the immediate environment around.

- User

  An employee who habitually uses DSE as a significant part of his normal work.

- Operator

  A self-employed person who habitually uses DSE as a significant part of his normal work.

Factors to consider in determining whether a person is a 'user' or 'operator' are:

- dependence on DSE to do the job;
- no discretion on whether to use the DSE;
- significant training or DSE skills required;
- use normally for continuous or near-continuous spells of an hour or more;
- use more or less daily;

- fast transfer of information is an important requirement;
- high attention and concentration required, where the consequences of an error may be critical.

Detailed examples are given in the HSE Guidance to the Regulations (**REF. 1**) providing pen-portraits of definite users, possible users and those who are definitely not users.

Exclusions

12.3 The Regulations do not apply to:

- drivers' cabs or control cabs for vehicles or machinery;
- DSE on board a means of transport;
- DSE mainly intended for public use;
- portable systems not in prolonged use;
- calculators, cash registers or other equipment with small displays;
- window typewriters.

Assessment of workstations and reduction of risk

12.4 *Regulation 2* requires employers to perform a 'suitable and sufficient' analysis of all workstations which are:

- used for their purposes by 'users';
- provided by them and used for their purpose by 'operators';

to assess the health and safety risks in consequence of that use. They must then reduce the risks identified to the lowest level reasonably practicable. As for other types of assessments, an assessment must be reviewed if there is reason to suspect it is no longer valid or there have been significant changes. Further guidance on assessment, including an assessment checklist, is provided later in the chapter.

Requirements for workstations

12.5 *Regulation 3* states that all workstations must meet the requirements laid down in the *Schedule* to the Regulations. The requirement must be relevant in relation to the health, safety and welfare of workers – for example, there is no need to provide a document holder (referred to in the *Schedule*) if there is little or no inputting from documents. The HSE guidance booklet (**REF. 1**) provides further examples of where the requirements of the *Schedule* may not be appropriate.

Daily work routine of users

12.6 Employers are required by *Regulation 4* to plan the activities of 'users' so that their DSE work is periodically interrupted by breaks or changes of activity. Breaks should be taken before the onset of fatigue and preferably away from the screen. It is best if users are given some discretion in planning their work and are able to arrange breaks informally rather than having formal breaks at regular intervals.

Eyes and eyesight

12.7 *Regulation 5* gives 'users' the right (at their employer's expense) to an appropriate eye and eyesight test by a competent person before becoming a 'user' and at regular intervals thereafter. (Some employers are re-testing at two yearly to five yearly intervals – in some cases varying the period depending on the age of the employee. Professional advice should be sought in cases of doubt.) They are also entitled to tests on experiencing visual difficulties which may reasonably be considered to be caused by DSE work. Such tests are an entitlement for 'users' and are not compulsory.

Tests are normally carried out by opticians and involve a test of vision and an examination of the eye. Where companies have vision screening facilities, 'users' may opt for a screening test to see if a full eye test is needed. If the eye test shows a need for spectacles (other than the 'user's' normal spectacles) the basic cost of these must be met by the employer, but employees must pay for extras, e.g. designer frames or tinted lenses.

Provision of training

12.8 'Users' are required by *Regulation* 6 to be provided with adequate health and safety training in the use of any workstation upon which they may be required to work. Training may also be required where workstations are substantially modified. The training should include :

- the causes of DSE-related problems, e.g. poor posture, screen reflections;
- the user's role in recognising risks;
- the importance of comfortable posture and postural change;
- equipment adjustment mechanisms, e.g. chairs, contrast, brightness;
- use and arrangement of workstation components;
- the need for regular screen cleaning;
- the need to take breaks and changes of activity;
- arrangements for reporting problems;
- information about the Regulations (especially sight tests and breaks);
- the user's role in assessments.

The HSE have published a leaflet (**REF. 2**) which provides a useful reference for training purposes.

Provision of information

12.9 *Regulation* 7 requires employers to ensure that operators and users at work within their undertaking are provided with adequate information. The table below shows the responsibility of the 'host' employer in this respect (and also gives a good guide to his responsibilities generally under the Regulations).

The 'host' employer must inform on:	Regulation:	Own users:	Other users (Agency Staff):	Operators (Self-employed):
DSE/Workstation Risks		YES	YES	YES
Risk Assessment and Reduction	2 and 3	YES	YES	YES
Breaks and Activity Changes	4	YES	YES	NO
Eye and Eyesight Tests	5	YES	NO	NO
Initial Training	6(1)	YES	NO	NO
Training when Workstation Substantially Modified	6(2)	YES	YES	NO

Planning and preparation

Who will carry out the assessments?

12.10 As with other types of assessments, DSE workstation assessments within an organisation may be carried out by an individual or by members of an assessment team. Those responsible for making assessments should have received appropriate training so that they are familiar with the requirements of the Regulations and they should have the ability to:

- assess risks from the workstation and the kind of work;
- use additional sources of information or expertise as appropriate (recognising their own limitations);
- draw valid and reliable conclusions;
- make a clear record and communicate the findings to those who need to take action.

They may be health and safety specialists, IT managers or line managers, and there will often be benefits in involving employees' safety representatives in the assessment process.

How will the assessments be organised?

12.11

Identify the 'users'

An important first step is to identify the 'users' (together with any 'operators') of DSE within the organisation. The definitions in *Regulation 1* refer to habitual use of DSE. The HSE guidance booklet (**REF. 1**) provides considerable advice on the factors which must be taken into account, of which time spent using DSE is the most significant. Some organisations have adopted a rule of thumb that anyone spending more than 50% of their time in DSE work is a DSE 'user'. However, the HSE guidance indicates that a less simplistic approach should be taken.

The assessment checklist provided later in the chapter includes reference to the other factors which the HSE state should be taken into account in deciding whether an individual is a 'user' or an 'operator'. It should be noted that employers have duties to assess workstations used for the purposes of their undertaking by *all* 'users' or 'operators'. This includes 'users' employed by others (e.g. agency-employed staff), 'operators' (e.g. self-employed draughtsmen or journalists), peripatetic staff (e.g. journalists, sales staff, careers advisors) and homeworkers.

In practice, most employers have not found it too difficult to decide who their 'users' and 'operators' are. Many have taken the approach of assessing the workstations of those individuals where doubt exists and, if necessary, making a final decision then. More time and expense can often be wasted debating a few borderline cases than would be involved in including them in the definition.

Decide on the assessment approach

Some of the author's practical experience of assessment work has involved his personal assessment of each individual 'user's' or 'operator's' workstation, using a simple checklist (a completed example of this checklist is provided later in the chapter). Many others have taken a similar approach to assessments. However, in some organisations the number of DSE workstations may be so large that this approach would either be impracticable or would require an unnecessarily large input of resources. In such cases the issuing of a self-assessment checklist to identified 'users' will often be more appropriate. A sample checklist, together with guidance to 'users' on its completion, is provided later in the chapter. The HSE guidance booklet (**REF. 1**) provides a checklist covering similar topics. Providing that workers are given the necessary training and guidance in how to use such checklists, this approach is perfectly acceptable. However, it must be supported by other actions such as:

- an inspection of areas where DSE workstations are situated (evaluating general issues such as lighting, blinds, housekeeping, desk space and the standards of chairs and DSE hardware);
- providing employees with the option of an assessment of their individual workstation by a specialist;
- responding promptly to problems identified in the completed checklists.

Making the assessments

Observations

12.12 Where workstations are to be assessed on an individual basis the assessment checklist questions are intended to identify the principal factors that the assessor(s) need to look out for (as illustrated in the completed example).

Some of the more common problems identified are related to:

- Posture:
  - height of screen;
  - height of seat or position of backrest;
  - position of keyboard or keyboard technique;
  - need for footrest or document holder.
- Vision:
  - angle of screen;
  - position of lights or need for diffusers;
  - need for blinds to control sunlight;
  - adjustment of brightness or contrast controls.

Discussions

12.13 One of the dangers of self-assessments is that a minority of employees will blame their DSE work for a variety of unrelated problems. An individual assessment provides an opportunity for a two-way dialogue between the assessor(s) and the 'user' and allows the assessor to evaluate whether stated problems are related to deficiencies in the workstation.

Common causes of problems are:

- Back, shoulders, neck;
  - height or position of screen;
  - positioning of seat;
  - need for footrest or document holder.
- Hands, wrists, arms;
  - position of keyboard;
  - keyboard technique.
- Tired eyes or headaches;
  - failure to take breaks or change activities;
  - reflected light (artificial or sunlight).
- Discussions with 'users' can also reveal other important pieces of information, e.g.;
  - there are problems with sunlight at certain times of day or periods of the year;
  - the 'user' does not know how to adjust their chair or brightness/contrast controls;
  - the 'user's' chair is broken and incapable of being adjusted;
  - the 'user' has never been offered an eye test.

Assessment records

12.14 As for other types of assessment, there is no standard format for DSE workstation assessment records. The completed sample assessment checklist and the self-assessment checklist (both below), together with the guidance on its completion are offered as examples of record formats that have been found successful in practice. (A six page workstation assessment checklist is also available from the HSE (**REF. 4**).)

The HSE guidance booklet (**REF. 1**) states that records may be stored in electronic as well as paper form. Self-assessment checklists would particularly lend themselves to being completed and submitted electronically. No guidance is provided on how long records should be kept. Prudent employers may prefer to retain them indefinitely, bearing in mind that civil claims for alleged DSE-related conditions may be submitted many years after it is claimed the condition was first initiated.

Display screen equipment workstation assessment

USER'S NAME: *Caroline Smith* LOCATION: *Accounts*

FACTOR			COMMENT
1	WORK PATTERNS		
1.1	Most time spent per day at DSE		*6 Hours*
1.2	Average time per day at DSE		*4–5 Hours*
1.3	Number of days per week at DSE		*5*
1.4	Longest spell without break		*1–2 Hours*
1.5	Can breaks be taken?		*YES – at Caroline's discretion*
1.6	Concentration important?		*Fairly Important – Accuracy*
1.7	Speed of operation important?		*NO*
'USER' STATUS CONFIRMED			*Yes*
2	PROBLEMS EXPERIENCED Has the user significant experience of problems with:		
2.1	Back, shoulders or neck		*Pains in shoulder and neck*
2.2	Hands wrists or arms		*None*
2.3	Tired eyes or headaches		*Occasionally*
2.4	Suitability of software		*No problems*
2.5	Other problems		*None*
3	LIGHTING/ENVIRONMENT		
3.1	Artificial Lighting:	Adequate to see documents	*YES*
		Any reflection or glare problems	*Some reflection from striplight (diffuser fitted)*
3.2	Sunlight:	Any reflection or glare problems	*In mornings from window behind.*
		Suitable blinds available (if necessary)	*Good vertical blinds provided*
3.3	Noise:	Hindering to communication	}*No problems*
		Distracting or stressful	}
3.4	Temperature & ventilation:	Satisfactory in summer and winter	*Office hot and stuffy in summer*
4	SCREEN		
4.1	Set at suitable height		*Screen too low*
4.2	Stable Image with clear characters		*Good. Able to vary colours*
4.3	Brightness and contrast adjustable		*Both have adjustable controls*
4.4	Swivels and tilts easily		*YES*
4.5	Cleaning materials available		*In stationery store*
5	KEYBOARD		
5.1	Separate and tiltable		*YES*
5.2	Sufficient space in front		*Too near edge of desk*
5.3	Keys clearly visible		*YES*

6	DESK AND CHAIR	
6.1	Desk size adequate	*Satisfactory*
6.2	Sufficient leg room	*Materials being stored in desk well*
6.3	Desk surface low reflectance	*YES – wooden surface*
6.4	Suitable document holder (if required)	*Not available*
6.5	Chair comfortable and stable	*YES*
6.6	Chair height adjustable	*YES*
6.7	Back adjustable (height and tilt)	*Adjustment not working*
6.8	Footrest available (if required)	*Not required*

OTHER COMMENTS

Caroline was advised to take short breaks more frequently.
Some lengthy jobs seem to be the cause of her tired eyes and headaches.
Repositioning of the screen and provision of a document holder should overcome the shoulder and neck problems.

No.	Actions required	Responsibility
3.1	Modify screen position to avoid light reflection.	*C Smith*
3.2	Close blinds when sunlight bright.	*C Smith*
3.4	Free up seized office windows.	*Office Manager*
4.1	Provide screen stand and keep top of screen at eye level.	*Office Manager/C Smith*
5.2	Keep wrist resting space in front of keyboard.	*C Smith*
6.2	Remove items from desk well.	*C Smith*
6.4	Provide document holder.	*Office Manager*
6.7	Repair chair back adjustment.	*Office Manager*

Assessor's name: *C Moore* Signature: *C Moore* Date: *14/5/99*

PROGRESS WITH ACTIONS

All recommendations acted upon although Caroline needs to remember to close the blinds on sunny days. No more problems with shoulders/neck or tired eyes/headaches.

C Moore 15/7/99

Planned date for assessment review: July 2001

	DISPLAY SCREEN EQUIPMENT WORKSTATION SELF ASSESSMENT CHECKLIST USER'S NAME:	LOCATION:
1	LIGHTING AND WORK ENVIRONMENT	COMMENTS
1.1	Is artificial lighting adequate?	
1.2	Does it cause any reflection or glare problems?	
1.3	Any reflection or glare problems from sunlight?	
1.4	Are suitable blinds available (if necessary)?	
1.5	Are temperature and ventilation satisfactory in summer and winter?	
2	SCREEN AND KEYBOARD	
2.1	Is your screen set at a suitable height?	
2.2	Stable image with clear characters?	
2.3	Brightness and contrast adjustable?	
2.4	Screen swivels and tilts easily?	
2.5	Cleaning materials available?	
2.6	Is your keyboard tiltable?	
2.7	Have you sufficient space in front of it?	
3	DESK AND CHAIR	
3.1	Is your desk size adequate?	
3.2	Is there sufficient leg room under it?	
3.3	Do you have a suitable document holder (if required)?	
3.4	Is your chair comfortable and stable?	
3.5	Can you adjust your seat height?	
3.6	Can you adjust the height and tilt of your chair back?	
3.7	Do you have a footrest (if required)?	
4	HAVE YOU HAD SIGNIFICANT EXPERIENCE OF PROBLEMS WITH:	
4.1	Your back, shoulders or neck?	
4.2	Your hands, wrists or arms?	
4.3	Tired eyes or headaches?	
4.4	The suitability of the software you use?	
4.5	Other problems?	

ANY OTHER PROBLEMS OR COMMENTS?

WOULD YOU LIKE YOUR WORKSTATION TO BE ASSESSED BY A SPECIALIST ? Yes/No

Signature: Date:

Guidance on completing the DSE workstation self-assessment

Lighting and work environment

1.1 Artificial lighting should be adequate to see all the documents you work with.

1.2 Recessed lights with diffusers shouldn't cause problems. Lights suspended from ceilings might.

1.3 There may be problems in the early morning or afternoon, especially in winter when the sun is low.

1.4 Blinds provided should be effective in eliminating glare from the sun.

1.5 Strong sunlight may create significant thermal gain at times.

Screen and keyboard

2.1 The top of your screen should normally be level with your eyes when you are sitting in a comfortable position.

2.2 There should be little or no flicker on your screen.

2.3 You should know where the brightness and contrast controls are.

2.4 The screen should swivel and tilt so that you can avoid reflections.

2.5 You should know where to get cleaning items for your screen (and keyboard, if necessary).

2.6 Small legs at the back of your keyboard should allow you to adjust its angle.

2.7 Space in front the keyboard allows you to rest your hands and wrists when not keying in.

Desk and chair

3.1 Your desk should have sufficient space to allow you to have your screen and keyboard in a comfortable position and accommodate documents, document holder, phone etc.

3.2 There should be enough space under the desk to allow you to move your legs freely.

3.3 If you are inputting from documents, using a document holder helps avoid frequent neck movements.

3.4 Chairs with castors must have at least five (four is very unstable).

3.5 You should be able to adjust your seat height to work in a comfortable position (arms approx. horizontal and eyes level with the top of the screen).

3.6 The angle and height of your back support should be adjustable so that it provides a comfortable working position.

3.7 DSE users who are shorter may need a footrest to help them keep comfortable when sitting at the right height for their keyboard and screen (see 3.5).

Possible problems

4.1 Problems with back, shoulders or neck might indicate your screen is at the wrong height, an incorrectly adjusted chair or need for a document holder.

4.2 Problems with hands, wrists or arms might indicate incorrect positioning of the keyboard or a poor keying technique. Hands should not be bent up at the wrist and a soft touch should be used on the keyboard, not overstretching the fingers.

4.3 Tired eyes or headaches could indicate problems with lighting, glare or reflections.

They may also indicate the need to take regular breaks away from the screen. Persistent problems might need an eye test – contact Human Resources to request one.

4.4 The software should be suitable for the work you have to do.

IF YOU HAVE A PROBLEM OR WOULD LIKE A SECOND OPINION, WE CAN ARRANGE FOR A SPECIALIST TO COME TO ASSESS YOUR WORKSTATION

Special situations

12.15 The revised HSE booklet (**REF. 1**) provides further guidance on several aspects of DSE work which have developed over the last decade as information technology has been transformed.

Homeworkers and teleworkers

12.16 The *DSE Regulations* still apply to users employed to work at home or elsewhere away from their main base, whether or not equipment is provided by the employer. It may not be possible to send someone to carry out a workstation assessment and here the HSE recommend the use of a self-assessment checklist (as provided earlier in this chapter or available in the HSE guidance booklet). Staff should receive appropriate training to carry out assessments, and responsibility must be clearly allocated for dealing with risks identified during the process.

Other issues to be emphasised during training are the importance of good posture and of taking breaks. Arrangements for reporting subsequent defects and problems should also be established. General aspects of homeworking, including electrical safety, are covered in a free HSE leaflet on the subject (**REF. 5**).

Shared workstations

12.17 'Hot desking' arrangements mean that many workstations are shared, sometimes by workers of widely differing sizes. Assessments of such workstations should take into account aspects such as:

- whether chair adjustments can accommodate all the workers involved;
- the availability of footrests;
- adequacy of legroom for taller workers;
- arrangements for adjusting the heights of screens (e.g. stands).

In some situations the use of desks with adjustable heights may be appropriate.

Work with portable DSE

12.18 Work with laptop computers as well as mobile phones or personal organisers which can be used to compose, edit or view text is becoming increasingly common. Where mobile phones or personal organisers are used in this way for prolonged periods they are subject to the Regulations, as is all work with laptops.

Mobile phones or personal organisers should not generally be used for sufficiently long periods to require a workstation assessment but work involving laptops is likely to be of much longer duration. The revised HSE booklet (**REF. 1**) contains considerable guidance on the selection and use of laptop computers.

The design of laptops is such that postural problems are much more likely to result and there will be a far greater need to ensure that sufficient breaks are taken. Where the laptop is used in the office, the risks can be reduced considerably by the use of a docking station and the HSE booklet also provides guidance on this. Alternatively a separate keyboard and monitor could be made available for working with the laptop in the office. Other risks associated with the use of portable equipment (e.g. manual handling issues and possible theft involving assault) should also be considered.

Pointing devices

12.19 Much work at DSE workstations involves the use of a mouse, trackball or similar device to move the cursor around the screen and carry out operations. The HSE booklet (**REF. 1**) provides guidance on:

- Choice of pointing devices

  Suitability of the device for:

  - the environment (space, position, dust, vibration);
  - the individual (right or left handed, physical limitations, existing upper limb disorder);
  - the task (some devices are better than others in respect of speed or accuracy).

- Use of pointing devices

  Issues to be considered include:

  - positioning (close to the midline of the user's body, not out to one side);
  - work surface (particularly its height and degree of support for the arm);
  - mousemats (smooth, large enough, without sharp edges);
  - software settings (suitable for the individual user);
  - task organisation (point device use mixed with other activities);
  - training (how to set up and use devices);
  - cleaning and maintenance.

The booklet also contains specific guidance on touch screens and speech interfaces.

After the assessment

Review and implementation of recommendations

12.20 The general guidance provided in **CHAPTER 4: CARRYING OUT RISK ASSESSMENTS** is equally applicable to recommendations made as a result of DSE workstation assessments. It should be noted that the sample assessment checklist includes a space which can be used at a follow-up of an assessment to describe 'Progress with actions'.

Assessment review/re-assessments

12.21 Changes to the layout of office accommodation often take place with bewildering rapidity. Some of these changes have significant implications for DSE workstations, others do not. To carry out a re-assessment every time the office layout is changed would be an inefficient use of resources.

A better approach is to make both management and DSE 'users' aware that a review or reassessment should be requested from the assessor or assessment team whenever significant changes take place. This can be supported by observations by the assessor(s) of changes which have occurred or are in progress.

Because of the frequency of changes a periodic review of DSE workstation assessments would be beneficial. The HSE do not provide any guidance on this aspect. Reviews at two yearly intervals may be appropriate for larger workplaces where changes are regularly taking place whilst five yearly intervals would probably be suitable for smaller workplaces where the layout of workstations is more static. Where self-assessments have been adopted, these also should be repeated both when users change their workstations and at regular intervals in any case.

References

HSE publications

12.22

1	L26	Display screen equipment work. *Health and Safety (Display Screen Equipment) Regulations* 1992. Guidance on Regulations (2003).
2	INDG 36	Working with VDUs (2003) – priced pack of leaflets.
3	HSG 90	The law on VDUs: An easy guide (2003)
4		VDU workstation checklist (2003)
5	INDG 226	Homeworking: Guidance for employers and employees on health and safety (1996)

13 Assessment of Personal Protective Equipment (PPE) requirements

In this chapter:

Introduction	**13.1**
The Regulations summarised	**13.2**
PPE defined	13.2
Assessment of PPE requirements	13.3
Maintenance, replacement and accommodation	13.4
Ensuring that PPE is used properly	13.5
PPE assessment in practice	**13.6**
What PPE may be required	13.6
Carrying out assessments	13.7
Assessment records	13.8
Sample PPE assessment records	**13.9–13.11**
Manufacture of pre-stressed concrete beams	13.9
Newspaper Printing	13.10
Erection of timber-framed structures on building sites	13.11
References	**13.12**

Introduction

13.1 The *Personal Protective Equipment at Work Regulations 1992* (the *PPE Regulations*) replaced much outdated law on PPE. Unlike many of the previous Regulations (which often only applied to limited processes or activities), these Regulations require an assessment of the PPE needs, whatever the work activity. If the assessment shows PPE is necessary then the employer must provide it, maintain it, instruct and train employees in its use, and take reasonable steps to ensure that they do use it. Some minor amendments were made to the Regulations by the *Health and Safety (Miscellaneous Amendments) Regulations 2002*. The self-employed are also covered by the Regulations.

However, several other sets of Regulations also contain important PPE requirements. They include:

- the *Control of Lead at Work Regulations 2002*;
- the *Ionising Radiations Regulations 1999*;
- the *Control of Asbestos at Work Regulations 2002*;

- the *Construction (Head Protection) Regulations 1989*;
- the *Control of Noise at Work Regulations 2005*;
- the *Control of Substances Hazardous to Health Regulations 2002;*
- the *Work at Height Regulations 2005.*

The Regulations summarised

PPE defined

13.2 PPE is defined in *Regulation 2* of the *PPE Regulations* as,

> all equipment (including clothing affording protection against the weather) which is intended to be worn or held by a person at work and which protects him against one or more risks to his health or safety.

This might include safety helmets, eye protection, safety footwear, gloves, high visibility clothing, etc. Waterproof, weatherproof or insulated clothing is subject to the Regulations if it is needed to protect employees against health and safety risks, but not otherwise.

Under *Regulation 3* certain types of equipment are excluded from the application of the Regulations, including ordinary working clothes and uniforms (not specifically giving protection), self-defence or deterrent equipment (e.g. panic alarms) and portable devices for detecting and signalling risks and nuisances (e.g. gas monitors).

Assessment of PPE requirements

13.3 Under *Regulation 3* of the *Management of Health and Safety at Work Regulations 1999* (the *Management Regulations*), employers must have made an assessment of the risks to their employees' health and safety whilst at work, in order to identify the most appropriate way of reducing those risks to an acceptable level. *Regulation 4* of the *PPE Regulations* requires that PPE is only chosen as the last resort, i.e. after such measures as elimination at source, engineering controls or safe systems of work. If the risk cannot be adequately controlled by such other means then the employer must provide employees with suitable PPE free of charge. This PPE must be readily available. Whilst much PPE is provided on a personal basis, in some circumstances PPE may be shared.

The Regulations refer to many factors which should be considered when selecting suitable PPE.

The PPE must:

- be appropriate for the risks, the conditions and the period for which it is worn;
- take account of ergonomic requirements, the wearer's state of health and the characteristics of the workstation;

- be capable of adjustment to fit the wearer correctly;
- prevent or adequately control the risk, without increasing overall risk;
- comply with appropriate standards (normally bearing the CE mark);
- be issued personally to individuals, where necessary for hygiene or health reasons;
- be compatible with other types of PPE being used (*Regulation 5* refers).

Regulation 6 requires the employer to ensure that an assessment is made in respect of PPE needs. This must include:

- identifying risks not avoided by other means;
- defining the characteristics required of the PPE;
- comparing these with the characteristics of PPE available;
- checking the compatibility of PPE worn simultaneously.

In simple and obvious cases the assessment need not be recorded. In more complex cases it must be recorded and kept readily available. It may be incorporated within the general risk assessment required by the *Management Regulations*. (Some of the sample assessment records in **CHAPTER 5: ASSESSMENT RECORDS** refer to PPE requirements.) Examples of specific PPE assessment records are also provided later in this chapter. Assessments must be reviewed if it is suspected they are no longer valid or in the event of significant changes taking place.

Maintenance, replacement and accommodation

13.4 *Regulation* 7 states that employers must ensure that PPE is maintained in an efficient state, efficient working order and in good repair. Depending upon the type of PPE and its circumstances of use, this might require specific arrangements to be made for its cleaning, disinfecting, inspection, examination, testing or repair. Responsibility for this maintenance should be clearly identified.

Some tasks could be given to the user but more complex work may require specially trained personnel. Records should be kept, where appropriate. In most cases it will be sufficient to follow manufacturers' instructions. Some types of PPE may need to be replaced periodically, e.g. many manufacturers recommend replacing safety helmets at least every five years.

Regulation 8 requires appropriate accommodation to be provided for PPE when it is not being used. This will vary according to the type of PPE – pegs (for helmets or weatherproof clothing), carrying cases (for safety spectacles), lockers, containers, etc. Storage accommodation may need to protect the PPE from contamination, damage or loss. Where PPE may become contaminated in use, it must be kept separate from other clothing and equipment.

Ensuring that PPE is used properly

13.5 *Regulation 9* states that employees must be provided with adequate and appropriate information, instruction and training on PPE they are required to use, including:

- why and when it must be used (this might include providing relevant signs);
- how to use it, and its limitations;
- arrangements for its maintenance and/or replacement.

For some PPE, correct fitting or adjustment may be necessary. Practical demonstrations may be appropriate for some types of PPE and repeat demonstrations may be required at regular intervals.

Employers

Employers have a clear duty under *Regulation 10* to take all reasonable steps to ensure that PPE is properly used by employees. This will normally involve taking a pro-active approach to policing standards of compliance.

Employees

Employees also have duties under *Regulations 10* and *11*:

- to use PPE in accordance with their training and instruction;
- to return PPE after use to the accommodation provided;
- to immediately report loss of or obvious defect in their PPE.

PPE assessment in practice

What PPE may be required

13.6 The HSE Guidance accompanying the Regulations (**REF. 1**) gives much detailed advice on what types of PPE are likely to be necessary to protect against many common types of risk. Further guidance is included in other HSE publications, some of which have been listed in the reference section of this chapter.

The main types to be considered are explained below.

Head protection

Safety helmets are necessary whenever there is significant risk from falling objects or impact with fixed objects. Other types of head protection which may be appropriate are bump caps (providing limited impact protection), caps or hairnets (protecting against scalping or entanglement) and crash helmets (for use on certain types of transport, e.g. all-terrain vehicles).

Eye protection

Activities and processes involving risks to the face and eyes include;

- contact with acids, alkalis and corrosive or irritant substances;
- work with power-driven tools creating chippings, etc.;
- work with molten metal and other molten substances;
- welding or burning operations and other activities producing intense light or other significant optical radiation;
- use of liquid, gas or vapour under pressure.

The type of eye protection chosen will depend on the nature of the risk and the circumstances of use. Types available include safety spectacles, goggles, eyeshields and visors.

Foot protection

Safety boots or shoes are necessary for any situation where there is risk from items falling onto the feet or from penetration through the soles or heels of footwear. They are likely to be required for construction, maintenance and warehouse work and heavier industrial processes. Work with hot metal is likely to require the use of foundry boots with quick release fastenings and without external features such as laces which could trap molten metal. Wellington boots may also be necessary for some activities or external locations.

Hand and arm protection

Gloves or gauntlets are necessary for protection against:

- cuts and abrasions;
- extremes of temperature;
- skin irritation and dermatitis;
- contact with toxic or corrosive liquids;
- molten metal.

The choice of glove material is important as it must be capable of protecting against the risk (or a combination of risks) but should also be comfortable for the wearer and compatible with the work being done.

Body protection

Types of body protection which may be necessary include:

- Overall and aprons – to protect against hazardous substances, welding spatter, etc.;
- Outdoor clothing – providing protection against cold, rain, etc.;
- High visibility clothing – for work on or close to roadways or mobile plant;
- Specialist clothing – e.g. for work with chain-saws;
- Life jackets or buoyancy aids – for work close to water;
- Harnesses or fall arrestors – for some types of work at height*;
- Molten metal clothing – for work at furnaces, etc. (this may also need to protect against radiant heat).

* The additional requirements of the *Work at Height Regulations* must also be taken into account – see **CHAPTER 16: ASSESSMENT OF WORK AT HEIGHT**.

Respiratory protection

See **CHAPTER 9: COSHH ASSESSMENTS** dealing with COSHH assessments.

Hearing protection

See **CHAPTER 10: NOISE ASSESSMENT** dealing with noise assessments.

(These last two types of PPE are not subject to the *PPE at Work Regulations* but are included for completeness.)

Once the risk has been identified through the assessment process there is a wide range of British Standards and European Standards which can be consulted to select the most appropriate type of PPE. Many suppliers will provide well-illustrated catalogues to aid in the selection process and their technical staff can provide more detailed guidance if required.

Carrying out assessments

13.7 Personal protective equipment needs should really be identified during other types of risk assessment – as described elsewhere in this book.

- General risk assessments – the need for head, eye, foot, hand/arm or body protection;
- Manual handling assessments – the need for hand/arm protection and also possibly for foot and body protection;

- COSHH assessments – the need for respiratory, hand/arm, eye and body protection and occasionally also foot protection;
- Noise assessments – the need for hearing protection;
- Work at height assessments – the need for harnesses, restraints and fall-arrestors.

However, some employers may prefer to carry out a separate review of their PPE requirements and arrangements in order to ensure that:

- no PPE needs have been overlooked;
- suitable PPE is available to meet those needs;
- employees are fully aware of PPE requirements and know how to obtain it and use it;
- suitable arrangements for maintaining and storing PPE are in place;
- signs are in place indicating PPE requirements, where appropriate;
- PPE requirements are actually being complied with.

It may also be the case that an employer becomes aware that other risk assessments are either inadequate or out of date. Such an employer may wish to carry out a review of PPE needs in the short term, prior to carrying out a more comprehensive and lengthier risk assessment programme.

Individuals or team members carrying out assessments of PPE needs are likely to be similar to those carrying out general risk assessments (as described in **CHAPTER 4: CARRYING OUT RISK ASSESSMENTS**). As well as having the qualities referred to in that chapter, they should also have a practical working knowledge of the requirements of the *PPE at Work Regulations*. Where PPE assessments are carried out separately the methodology will be similar to that described for other types of assessments, i.e.:

- planning and preparation (including reference to HSE material and other relevant sources of information);
- assessment in the workplace (including observations and discussions with workpeople);
- the review and implementation of recommendations;
- subsequent review of the assessments (if suspected to be no longer valid or in the event of significant change).

Assessment records

13.8 As stated earlier and demonstrated in **CHAPTER 5: ASSESSMENT RECORDS** and elsewhere, PPE requirements can be incorporated within general risk assessment records or the records kept in relation to more specific risk assessment requirements (e.g. manual handling, COSHH, noise). The last sample record included in **CHAPTER 5: ASSESSMENT RECORDS** shows how PPE requirements can be included in comprehensive operating procedures, which also detail other health and safety precautions that are required.

However, many employers also choose to list PPE requirements separately for a variety of reasons which include:

- as an aid in communicating PPE requirements to employees (particularly during induction training);
- as a clear reference point to assist in the enforcement of PPE rules;
- as a demonstration to others (HSE Inspectors, clients, main contractors) that the employer has carried out a thorough PPE assessment.

Some examples of this type of PPE assessment record follow. These relate to:

- the manufacture of pre-stressed concrete beams (the same process referred to in the final example in **CHAPTER 5: ASSESSMENT RECORDS**);
- maintenance activities in a newspaper printing plant;
- the erection of timber-framed structures on building sites.

In each of these examples no differentiation has been made between PPE required under the *PPE at Work Regulations* and that required to comply with Manual Handling Operations, COSHH or *Noise at Work Regulations*.

Sample PPE assessment records

Manufacture of pre-stressed concrete beams

Personal Protective Equipment Requirements

13.9

TYPE OF PPE	MUST BE USED IN THESE LOCATIONS OR FOR THESE ACTIVITIES
Safety footwear:	
Safety boots, shoes or Wellington boots	By all operatives at all times. By anyone else directly involved in the manufacturing process or the stacking of beams.
Eye protection:	
Face visor or goggles	Using a Stihl saw or electric saw. Cutting wires under tension. Using an airline to clean the moulds.
Hearing protection:	
Ear muffs or ear plugs	Using a Stihl saw or electric saw to cut wires or concrete. (and by anyone working in the immediate vicinity) Using an airline to clean the moulds. (Use of hearing protection is recommended when the vibrating beam is operating nearby.)
Dust masks:	
Disposable masks for nuisance dusts	Using a Stihl saw or electric saw to cut out spacers or to cut beams.
Gloves:	
	Using a Stihl saw or electric saw. Placing wooden spacers under or on top of beams. Turning beams. Repositioning beams on the fork lift truck. Spraying moulds with releasing oil. Pulling the locator bar. Raking, cleaning and smoothing concrete.

Newspaper Printing

Personal Protective Equipment Requirements

13.10

Section: Maintenance	
PPE normally required (to be worn at all times when working in operational areas):	
• Safety footwear • Two piece suit	
PPE required for specific activities or locations:	
• When using most power tools or workshop machinery (inc. grinding, drilling, milling, turning):	Eye protection (goggles or safety spectacles)
• When blowing down with compressed air:	Eye protection (goggles or safety spectacles) Dust mask (if significant dust likely)
• When burning or welding:	Visor or goggles (with appropriate shade of filter) Leather gauntlets or gloves
• When working on the press solvent system:	Solvent resistant gloves (at all times) *Eye protection (goggles or safety spectacles) *Protective suits (unless exposure to solvent is minimal) *Respirators suitable for solvent vapours (*unless exposure to solvent is minimal)
• When working on the press fount spray system	Chemical resistant gloves are recommended
• When handling rough materials or equipment	General purpose gloves
• When working in designated hearing protection zones	Hearing protection (muffs, plugs on bands, disposable plugs)
• EMPLOYEES MUST: – use PPE as designated above – take care of PPE which is in their care – obtain replacements of PPE which has been damaged or lost – report to management any difficulties in obtaining suitable PPE.	

Erection of timber-framed structures on building sites

Personal protective equipment standards

13.11 All staff carrying out erecting work on site must conform to the PPE Standards set out in the table below. Specific activities or specific sites may require additional PPE to be provided and used as necessary.

TYPE OF PPE	REQUIREMENTS
Safety footwear	At all times when working on sites.
Safety helmet	At all times when working on sites.
Eye protection (safety glasses or goggles)	When using nail guns or powered woodworking machines. When using timber treatment chemicals.
Hearing protection	When using nail guns or powered woodworking machines.
Gloves (a suitable type)	When handling rough materials. When using timber treatment chemicals.
Waterproof clothing	If work is carried out in rain, hail or snow.
High visibility clothing	Work in the vicinity of lifting operations. If required by the Principal Contractor or Client.
Harnesses plus restraints and/or fall-arrestors	Work at height in areas not protected by suitable guard rails and where air bags are not in use.

Staff supervising site activities must ensure that all staff comply with these standards. (Individual employees and subcontractors also have legal duties to comply with PPE requirements.)

References

(HSE publications)

13.12

1	L 25	Personal Protective Equipment at Work *Personal Protective Equipment at Work Regulations 1992.* Guidance on Regulations (2005)
2	INDG 174	A Short Guide to the *Personal Protective Equipment at Work Regulations 1992*(2005) – free leaflet
3	APIS 1	PPE, High visibility clothing for airport workers (1995) – free leaflet
4	HSG 150	Health and Safety in Construction (2001) – pages 82–83
5	INDG 337	Sun protection. Advice for employers of outdoor workers (2001) – free leaflet
6	HSG 206	Cost and effectiveness of chemical protective gloves for the workplace (2001)
7	INDG 330	Selecting protective gloves for work with chemicals (2000) – free leaflet
8		The selection, use and maintenance of molten metal protective clothing (1996)

14 Fire and DSEAR assessments

In this chapter:

Introduction	**14.1**
Regulatory Reform (Fire Safety) Order 2005	**14.2**
Enforcement	**14.2**
Important definitions and duties	**14.3**
Fire safety duties	**14.4**
Guidance on the Order	**14.5**
Risk assessment under the Order	**14.6**
Planning and preparation for fire risk assessments	**14.7**
Who will carry out the assessment?	**14.7**
How will the assessment be organised?	**14.8**
Gathering information	**14.9**
Fire risk assessment factors	**14.10**
Fire detection	**14.11**
Fire alarms	**14.12**
Fire escape routes	**14.13**
Fire procedures	**14.14**
Fire fighting	**14.15**
Fire prevention	**14.16**
Fire-related signs	**14.17**
Making the assessment	**14.18**
Observations	**14.18**
Discussions	**14.19**
Checks of records	**14.20**
Assessment records	**14.21**
Sample fire risk assessment record	**14.22**
After the assessment	**14.23**
Review and implementation of recommendations	**14.23**
Maintaining fire precautions	**14.24**
Assessment review	**14.25**
Dangerous Substances and Explosive Atmospheres Regulations 2002 (DSEAR)	**14.26**
Dangerous substances	**14.27**
Explosive atmosphere	**14.28**
Main requirements of the Regulations	**14.29**
Control measures	**14.30**
DSEAR assessment records	**14.31**
Sample DSEAR assessment record	**14.32**
References	**14.33**

Introduction

14.1 From 1 October 2006, fire safety within the workplace (and many other sectors) is due to be covered by the *Regulatory Reform (Fire Safety) Order 2005*. For brevity this Order (which only applies in England and Wales) will be referred to in this chapter as the Order or the Fire Safety Order.

Whilst it is true that the Order simplifies the previous legal position, with 53 articles and 5 schedules it is still extremely lengthy and complex. In a number of places it duplicates requirements which are already present in health and safety legislation, particularly in the *Management Regulations 1999*.

Over 40% of the Order consists of amendments, repeals and revocations of previous requirements. The major changes affecting the workplace are that the Fire Safety Order will replace the requirement for Fire Certificates contained in the *Fire Precautions Act 1971* and also replace the *Fire Precautions (Workplace) Regulations 1997*, as amended in 1999. References to these latter regulations in the Management Regulations will also be removed.

Regulatory Reform (Fire Safety) Order 2005

Enforcement

14.2 The enforcing authority in respect of most premises covered by the Order is the fire and rescue authority for the area. However, the HSE are the enforcing authority for some premises subject to the *Nuclear Installations Act 1965*, work on naval ships and, in particular, construction sites. Other enforcing arrangements also exist for sports grounds (local authorities enforce), Crown premises, UKAE premises and military establishments.

Powers granted to inspectors under the Order are similar to those given under *HSWA 1974*. Enforcing authorities may issue Alterations Notices, Enforcement Notices and Prohibition Notices. Before passing plans under building regulations in respect of new buildings, extensions, structural alterations or change of use of buildings, local authorities must consult the enforcing authority under the Order.

Important definitions and duties

14.3 Part 1 of the Order provides the meanings of many terms used elsewhere in the Order.

The 'responsible person' is defined in *Article 3* as

(a) in relation to a workplace, the employer, if the workplace is to any extent under his control;

(b) in relation to any premises not falling within paragraph (a) –

(i) the person who has control of the premises (as occupier or otherwise) in connection with the carrying on by him of a trade, business or undertaking (for profit or not); or

(ii) the owner, where the person in control of the premises does not have control in connection with the carrying on by that person of a trade, business or undertaking.

(Consequently for workplaces the employer will normally be the 'responsible person'.)

The term 'general fire precautions' is defined in *Article 4*. This can be summarised as

- measures to reduce the risk of fire and spread of fire;
- means of escape and their safe and effective use;
- measures for fighting fires;
- means for detecting fire and giving warning;
- arrangements for actions in the event of fire (including instruction, training and mitigation measures).

Article 5 primarily places on the 'responsible person' the duties imposed by *Articles 8–22* of the Order, and any regulations made under the Order. However, the article also places some duties on those who, by contractual or tenancy agreements, have obligations in relation to the maintenance, repair or safety of premises.

By virtue of *Article 6*, the Order does not apply in relation to domestic premises; offshore installations; fields, woods or other land on agricultural or forestry undertakings; mines; borehole sites; and certain types of transport. *Article* 7 also contains other disapplications.

Fire safety duties

14.4 *Part 2* of the Order contains the various duties which must be carried out by the 'responsible person'. *Article 8* contains a 'Duty to take general fire precautions' under which the 'responsible person' must –

(a) take such general fire precautions as will ensure, so far as is reasonably practicable, the safety of any of his employees; and

(b) in relation to relevant persons who are not his employees, take such general fire precautions as may reasonably be required in the circumstances of the case to ensure that the premises are safe.

(The definition of 'relevant persons' includes persons lawfully on the premises and those in the immediate vicinity who are at risk.)

Article 9 requires the 'responsible person' to make a risk assessment – this is dealt with in 14.6 below. This requirement and several others in this part of the Order match closely those contained in the *Management Regulations 1999*. These include the following articles

10 Principles of prevention to be applied.
11 Fire safety arrangements.
18 Safety assistance.
19 Provision of information to employees.
20 Provision of information to employers and the self-employed from outside undertakings.
21 Training.
22 Co-operation and co-ordination.
23 General duties of employees at work.

Articles 13–17 relate to various specific aspects of fire precautions. These and several of the above requirements are dealt with in more detail in this chapter.

Article 12 requires the 'Elimination or reduction of risks from dangerous substances'. This is similar to requirements contained in DSEAR. These regulations are explained more fully from 14.26 onwards in this chapter.

Guidance on the Order

14.5 A series of guides are to be produced to provide detailed information on risk assessments and other issues. These deal with

1. Offices and Shops.
2. Premises providing Sleeping Accommodation.
3. Residential Care.
4. Small and Medium Places of Assembly.
5. Large Places of Assembly.
6. Factories and Warehouses.
7. Theatres and Cinemas.
8. Educational Premises.
9. Healthcare Premises.
10. Transport Premises and Facilities.
11. Open Air Events.

These, together with a small entry level guide to the Order, should be available via the websites of the Office of the Deputy Prime Minister and the Small Business Service of Business Link. It has not been made clear whether *Fire Safety. An employer's guide* (a joint Home Office/HSE publication) is to be replaced by the above guides, or is to be revised to take account of the Order. Nevertheless, even if out of date, this booklet still contains much practical information of assistance in carrying out a fire risk assessment.

Risk assessment under the Order

14.6 *Paragraph (1)* of *Article 9* states that:

> The responsible person must make a suitable and sufficient assessment of the risks to which relevant persons are exposed for the purpose of identifying the general fire precautions he needs to take to comply with the requirements and prohibitions imposed on him by or under this Order.

Subsequent parts of *Article 9* contain additional requirements relating to the risk assessment. It must:

- include consideration of dangerous substances (*Part 1* of *Schedule 1* to the Order contains detailed requirements);
- be reviewed regularly to keep it up to date (particularly if it is suspected to be no longer valid or significant changes occur);
- take account of young persons employed;
- be recorded if five or more persons are employed (or if a licence or alterations notice is in force for the premises).

No new work activity involving a dangerous substance may commence unless a risk assessment has been made and measures required under the Order have been implemented.

Planning and preparation for fire risk assessments

Who will carry out the assessment?

14.7 *Article 18* of the Fire Safety Order requires the appointment of 'one or more competent persons' to provide safety assistance. The detailed requirements of this article are similar to those for health and safety assistance contained in *Regulation 7* of the *Management Regulations 1999* (see **2.6: WHO SHOULD CARRY OUT THE ASSESSMENT**). The qualities required by assessors will depend upon the size, complexity and fire risks of the workplace concerned. Employers, managers or safety representatives without any specialist knowledge should be capable of dealing with small, low risk situations but more complex, higher risk premises will require a greater degree of knowledge and experience.

How will the assessment be organised?

14.8 As for other types of assessment, it may be appropriate to divide the workplace into assessment units. However, the size of the units is likely to be larger than for general risk assessments. This is partly because only fire issues need to be considered as opposed to a multiplicity of types of risk. This is also likely as fire protection systems (fire detection, fire alarms, sprinklers, evacuation procedures) are usually integrated within large buildings.

However, there is no reason why fire issues cannot be dealt with during general risk assessments in each assessment unit (as described in **CHAPTER 4: CARRYING OUT RISK ASSESSMENTS**), with the fire protection systems then being assessed separately for the premises as a whole. If workplaces are divided into units purely for fire assessment, then the division may be on the basis of:

- separate buildings;
- major sections of buildings, e.g. operational departments;
- separate floors.

It is also important that fire risks in all parts of the workplace are properly assessed. Areas which could be overlooked are:

- outbuildings such as boiler-houses or stores;
- basements or pits below equipment;
- overhead crane and similar control cabs;
- upper walkways and platforms;
- roof areas where provision is made for maintenance access;
- plant rooms;
- confined spaces to which access is made periodically.

Assessments must also be made in respect of construction sites for which the HSE are the enforcing authority (see **REF. 2**).

Where the workplace is in a shared building there will need to be liaison with other occupants and/or the landlord or managing agent. This liaison is particularly important where the landlord, the managing agent or a major occupier is responsible for fire protection systems, e.g. fire detection or fire alarms. Such persons also have responsibilities under the Fire Safety Order.

Gathering information

14.9

Fire certificate

Even though Fire Certificates will no longer be a legal requirement, a reasonably up-to-date certificate will still be extremely useful in carrying out a risk assessment. Assessors should be able to have some confidence that if a fully trained fire officer has considered the fire evacuation routes and fire protection systems to be good enough to issue the certificate, then they met relevant standards at the time of issue. However, the existence of the Fire Certificate only provides a valuable starting point. The assessor(s) will still need to:

- look for changes, e.g. to premises, activities, materials used or stored;
- ensure that fire protection equipment is being inspected, maintained, tested (the Fire Certificate will usually have specified standards for this); and
- evaluate the standards of fire prevention within the workplace.

In practice many Fire Certificates have been allowed to become considerably out of date due to the failure of employers to notify changes or failure of fire authorities to react to such notifications.

Plans and drawings

Where there is no Fire Certificate or the certificate is out of date, plans and drawings of the premises are extremely useful in carrying out the fire risk assessment. Even simple layout plans can be marked up to show the fire evacuation routes and the positions of detectors, alarm call points, alarm sounders, fire fighting equipment, etc. These marked up drawings can form an important part of the eventual assessment records. Detailed drawings which already show the positions of such equipment are a bonus (although the assessor(s) should always be alert for possible inaccuracies).

Records

Records of maintenance, tests and inspections of fire protection systems will need to be examined as part of the assessment. These could be checked prior to or during the assessment. Such records might include those relating to:

- fire detection equipment (smoke and heat alarms);
- fire alarms;
- emergency lighting;
- fire fighting equipment (extinguishers, hose reels, sprinklers);
- fire evacuation drills.

Reference material

Guidance available on the Order was referred to in **14.5** above. There are many other sources of reference in respect of different types of premises, specialist fire risks or particular types of fire precautions. A range of these are listed in the reference section of this chapter but far more are available.

Fire risk assessment factors

14.10 Detailed guidance is available from various sources (including those referred to in **14.5**) on the various factors which should be taken into account when carrying out a fire risk assessment. It is not possible to provide the same level of detail in a book of this nature but this section includes reference to the principles to be followed and some of the key standards relating to:

- Fire detection
- Fire alarms
- Fire escape routes
- Fire procedures
- Fire fighting
- Fire prevention
- Fire-related signs.

Fire detection

14.11 *Paragraph (1)* of *Article 13* of the Order states 'Where necessary (whether due to the features of the premises, the activity carried on there, any hazard present or any other relevant circumstances) in order to safeguard the safety of relevant persons, the responsible person must ensure that –

(a) the premises are, to the extent that it is appropriate, equipped with appropriate fire-fighting equipment and with fire detectors and alarms; . . . '

However, this does not mean that every workplace must have automatic fire detection equipment. In many workplaces fires can be detected during working hours by observation and smell.

During the assessment, consideration must be given to the possibility of fires developing in sleeping accommodation (e.g. in hotels, hostels, care homes) or starting in unoccupied areas of other premises and spreading. Even during working hours fires could develop undetected:

- in storage areas or storerooms;
- in plant rooms;
- in thinly populated work areas;
- during break times.

If such a fire could significantly endanger the safety of employees and other relevant persons (eg by cutting off a single escape route) then some sort of fire detection equipment must be provided. Fire detectors may also be justified in order to prevent or reduce property damage – many insurers will insist upon its provision in certain areas.

The nature of any fire detection equipment will vary according to the extent and type of fire risk present.

Domestic smoke alarms

Where the only escape route from a work area (e.g. an office or workshop) is through an unoccupied outer area (e.g. a storeroom), the provision of a single domestic smoke alarm may be appropriate. Such an alarm should conform with British Standard 5446: *'Components of automatic fire alarm systems for residential premises. Part 1: Specifications for self-contained smoke alarms and point-type smoke detectors'* (**REF. 3**).

Interlinked systems

Similarly such domestic smoke alarms can be interlinked to provide protection in slightly more complex situations. However, higher risk workplaces will require more sophisticated and reliable detection systems conforming with British Standard 5839: *'Fire detection and alarm systems for buildings Part 1'* (**REF. 4**).

Types of detectors

Basic domestic smoke detectors are usually more sensitive than those in more sophisticated systems, leading to a greater possibility of false alarms. In some workplaces it may be better to install heat detectors because smoke, fumes or dust from work activities would trigger off any type of smoke detector. Specialist advice should be sought before installing any automatic fire detection system or an interlinked system of smoke alarms. Most fire authorities are willing to provide such advice.

Fire alarms

14.12 *Article 13(1)(a)* (see 14.11 above) also requires consideration to be given to the need for fire alarms. As for fire detection, the degree of sophistication should match the size of the premises and the extent of the fire risk. In many small buildings or larger but open rooms, a shout of 'Fire' can be heard by everyone, including those who may be in restrooms, toilets or storerooms.

Slightly larger premises may justify the installation of a single combined alarm point including a call point, bell and battery (plus charger). In both cases the unit should be positioned so that it can be reached quickly and operated without exposing the person using it to danger.

Premises of any significant size will require a conventional type of electrical fire alarm system incorporating a number of manually activated call points (usually situated at or close to principal exits from the building or sections of it). The system should have sufficient sounders for the alarm to be heard throughout the premises. In noisy premises sounders may need to be augmented by visual alarms, e.g. distinctive flashing lights. (These may also be necessary where occupants have hearing problems.) Such systems are normally integrated with any automatic fire detection equipment which may have been installed – British Standard 5839: *Part 1* provides a Code of Practice for design, installation and servicing (**REF. 4**).

Some premises use a two tier system of alarms. One type of sound indicates a fire problem which must be investigated whilst a different sound requires the premises to be evacuated. Some types of premises (e.g. large retail operations or places of entertainment) utilise public address systems as part of their fire alarm system. This can provide guidance on the action to be taken – important where many of those present will be unfamiliar with the building and evacuation procedures. BS 5839: *Part 8* provides a Code of Practice for such systems (**REF. 4**).

Fire escape routes

14.13 *Article 14* of the Order sets out a number of requirements for emergency routes and exits. It must be possible for persons to evacuate the premises as quickly and as safely as possible. Routes and exits must be adequate for the use, equipment and dimensions of the premises and for the maximum number of persons who may be present there at any one time. It will usually be necessary to refer to relevant guidance (see 14.5) to ensure that the premises meet the appropriate standards.

Whilst having more than one route to escape is always preferable, it is not possible in every case. There will be many parts of premises (individual offices, storerooms, kitchens, toilets, bedrooms) which only have one escape route.

Several other requirements are also contained in *Article 14*. Where necessary:

- emergency routes and exits must be kept clear at all times;
- routes and exits must lead as directly as possible to a place of safety;
- emergency doors must open in the direction of travel;
- sliding or revolving doors must not be used specifically as emergency exits;

– emergency doors must not be locked or fastened so that they cannot be easily and immediately opened in an emergency;

– emergency routes and exits must be indicated by signs;

– routes and exits requiring illumination must be provided with adequate emergency lighting.

Use of the phrase 'where necessary' means that these are not absolute requirements. It is the purpose of the assessment (taking into account relevant guidance) to interpret what is necessary in the premises being assessed.

The maintenance of protected fire escape routes is extremely important – leaving self-closing fire doors wedged open can result in fires and products of combustion (particularly smoke) spreading rapidly throughout buildings.

Fire procedures

14.14 *Article 15* of the Order requires the establishment of appropriate fire procedures, including safety drills. These procedures must include:

– sufficient competent persons to implement evacuation from the premises;

– restricting access to high risk areas.

Paragraph (3) of *Article 13* also requires, where necessary:

– measures for fire-fighting in the premises (and persons to implement these measures);

– contacts with external emergency services (for fire-fighting, rescue work, first aid and emergency medical care).

Fire procedures would normally contain the following elements:

Action on finding a fire

This can be done usually setting off the alarm or shouting 'Fire'. Employees may be invited to tackle the fire, if they can do this without endangering themselves.

Reaction to the fire alarm

Normally this would be to evacuate using the nearest safe exit route. Some employees may be designated to carry out specific tasks, e.g.:

- telephoning the Fire Brigade and any other relevant emergency services (if the alarm does not do this automatically);
- collecting visitor books (or similar records);

- carrying out sweeps to ensure that all persons including non-employees (e.g. guests or customers) have evacuated designated areas;
- carrying out emergency shutdowns of equipment, if they can do this without risk;
- carrying out other emergency actions, including fire fighting.

A designated assembly point

In large premises different groups may be allocated separate assembly points. At the assembly point a roll call should be carried out to establish whether everyone can be accounted for. Such roll calls would be carried out by fire wardens (each of whom would have a designated deputy). Some wardens may have been given responsibilities to 'sweep' designated areas. Administrative staff (who are aware of the movements of employees and others) are often more suitable for this post than more senior staff, who may not always be on the premises.

Fire wardens are expected to be the point of contact with the Fire Brigade and other emergency services, especially in respect of persons who cannot be accounted for and when it is safe for premises can be re-entered.

Regular fire evacuation drills are an essential part of ensuring that the evacuation procedure will be effective if it is really needed. Drill frequencies were often previously specified in Fire Certificates. They should normally be held at intervals of 3–12 months, depending upon the level of fire risk. Records should be maintained of the dates of the drills, the time taken to account for the safe evacuation of the premises and any other relevant details.

Training of staff in the fire evacuation procedure should be a key element of induction. Arrangements should also be in place to inform contractors and others on the premises of fire procedures. Further training will usually be necessary for those with specific roles in the event of fire, e.g. fire wardens. All staff are likely to need to be reminded periodically about fire procedures.

Fire fighting

14.15 *Paragraph (1)* of *Article 13* of the Order requires the provision of fire fighting equipment. Where necessary, non-automatic equipment must be accessible, simple to use and indicated by signs. The assessment must determine what fire fighting equipment is necessary. This may include:

- portable fire extinguishers (see below);
- fire blankets;
- hose reels;

- sprinkler systems;
- other fixed fire fighting systems (utilising carbon dioxide or other inert gas, foam, dry powder).

Extinguishers are usually best located in positions from which those using them can, if necessary, make a safe escape, i.e. close to fire exit doors or on fire exit routes. Those hidden from view (e.g. in alcoves or containers) may need to be indicated by signs, but signs need only be provided where necessary.

Where staff are specifically given fire fighting roles in the event of a fire, they must be given adequate training. It is also desirable for other staff to be given basic training, e.g. on the suitability of different extinguishers for different types of fires (see below) and preferably also in practical fire fighting techniques.

The Regulations on safety signs and signals now require fire extinguishers to be predominantly red, rather than each type being coloured differently, as was previously the case. However, extinguishers in the UK are still provided with additional coloured labels to denote the different types, in accordance with British Standard 7863: '*Recommendations for colour coding to indicate the extinguishing media contained in portable fire extinguishers*' (**REF. 5**). British Standard 5423: '*Portable fire extinguishers*' provides detailed guidance on portable fire extinguishers (**REF. 6**).

EXTINGUISHER TYPE	Colour Code	Suitability
WATER	Red	Wood, paper, textiles NOT electrical or flammable liquid fires
FOAM	Cream	Flammable liquids Reasonable for wood, paper, textiles NOT electrical fires
CARBON DIOXIDE	Black	Flammable liquids and best for electrical fires NOT paper fires
DRY POWDER	Blue	Flammable liquids, electrical fires Reasonable for wood, paper, textiles
VAPOURISING LIQUID	Green	Flammable liquids and electrical fires
FIRE BLANKET	–	Flammable liquids in containers e.g. deep fat friers, chip pans

Fire prevention

14.16 Fire prevention is an important part of fire precautions and ensuring the safety of employees and other relevant persons, as required by *Article 8* of the Order. The risk assessment must take account of precautionary measures such as:

- safe storage of all dangerous substances, particularly flammable liquids and gases;
- minimising quantities of dangerous substances present in premises;

- safe methods of use of dangerous substances;
- maintenance of plant and equipment (avoiding electrical short circuits, overheating);
- good standards of housekeeping;
- control of hot work (burning, welding, etc.);
- careful positioning of heating appliances;
- control of smoking;
- control and removal of flammable rubbish and waste;
- controlled burning of rubbish;
- checks before locking up and/or regular patrols by security staff;
- other measures to avoid arson.

Fire-related signs

14.17 The need for fire-related signs may be identified during the assessment although such signs are *not* an automatic requirement, they need only be provided where necessary. Any signs which are provided must comply with the requirements of the *Health and Safety (Safety Signs and Signals) Regulations 1996* (**REF. 7**).

Situations where signs may be required are given below.

Fire alarm call points

Where call points may not be immediately obvious, particularly if they may need to be operated by people not familiar with the premises.

Fire escape routes

Exit routes and doors not in regular use or not otherwise obvious (again taking into account persons not familiar with the building). All such signs must now include a pictogram symbol.

Evacuation procedures

Fire Action Notices can be used to summarise evacuation procedures. Some types of sign allow the insertion of local details, e.g. the fire assembly point. Such signs should be in addition to, not in place of, the provision of staff training.

Fire fighting equipment

Signs may be used to indicate the positions of fire fighting equipment if this is not readily apparent.

Fire prevention

Fire prevention requirements, e.g. no smoking, no hot work, may be emphasised by signs.

Many types of fire sign are now available in luminous materials and fire exit route signs can be incorporated into emergency lighting installations.

Making the assessment

Observations

14.18 Observations during a fire risk assessment should relate closely to the assessment factors described above. Aspects to particularly look out for will include:

- areas where fire might break out undetected;
- escape routes through unoccupied areas;
- positioning of fire alarm call points;
- single escape route situations;
- obstructed means of escape;
- wedged self-closing fire doors;
- easy opening of fire exit doors;
- signs indicating means of escape;
- display of information re-evacuation procedures;
- positioning (and possible obstruction) of fire fighting equipment;
- condition of fire extinguishers and evidence of their maintenance;
- storage and use of dangerous substances;
- housekeeping standards;
- other fire prevention measures.

Discussions

14.19 Questions to workpeople during the assessment will also relate to the assessment factors and might include:

- audibility of fire alarms;
- evidence of fire alarm tests;
- knowledge of evacuation routes and assembly point;
- awareness of roll-call arrangements, etc.;
- arrangements for dealing with non-employees (e.g. residents, customers, service users, visitors)
- evidence of evacuation drills;
- fire-related training (at induction and other times);
- knowledge of safe systems of work relating to dangerous substances or potential sources of ignition.

Checks of records

14.20 Records likely to be examined during the assessment (and where appropriate related to standards specified in the Fire Certificate) will include those of the testing and maintenance of:

- fire detection equipment;
- fire alarms;
- emergency lighting;
- fire fighting equipment.

Checks should also be made on records of:

- fire evacuation training and drills;
- induction training;
- fire fighting training;
- other emergency-related training.

Assessment records

14.21 As stated in **14.6**, all employers with five or more employees must record the significant findings of their risk assessment. As with other types of assessment, there is no standard format for recording fire risk assessments. However, some enforcing authorities may strongly encourage the use of particular styles of assessment record. The risk assessment forms demonstrated in **CHAPTER 5: ASSESSMENT RECORDS** can quite easily be used for recording this type of risk assessment.

The fire risk assessment factors described earlier in the chapter can be utilised as headings within the assessment record. Plans showing the layout of the workplace, including the location of any significant fire risks and the positioning of fire protection equipment, can be incorporated into the record. However, cross reference can also be made from the assessment record to a previous Fire Certificate, providing any significant changes are identified.

The headings below have been used in the sample fire risk assessment record provided in the chapter for a vehicle body repair business:

1 Introduction.
2 Summary (inc. reference to drawings, plans or a Fire Certificate).
3 Fire risks and their control.
4 Fire detection.
5 Fire alarm.
6 Means of escape (including provision of signs and emergency lighting).
7 Fire evacuation arrangements.
8 Fire fighting equipment.
9 Recommendations:
    - Appendices (drawings detailing)
        - Escape routes and fire alarm call points,
        - Smoke and heat detectors,
        - Fire fighting equipment.

Sample fire risk assessment record

14.22

BODY BEAUTIFUL REPAIR CENTRE

FIRE RISK ASSESSMENT

1. **Introduction**

The centre carries out repairs to the bodywork of cars and light vans (any mechanical work required is done at an associated workshop nearby). The main part of the premises consists of a large workshop containing two vehicle finishing units, separated by a small paint preparation room. Body preparation work is carried out in other parts of the workshop. There is a small parts department on two levels on the eastern side of the workshop.

Office accommodation is on two floors at the front (south side) of the premises. The ground floor of the offices communicates directly with the workshop. There is a large enclosed compound at the rear of the premises. On the western side of the compound is a small building used to store flammable liquids.

2. **Summary**

A Fire Certificate (not now a legal requirement) was issued by the local Fire Authority for the premises on 8 January 2002 and no significant changes have been made since that date. Several references are made in this assessment to the drawings accompanying the Fire Certificate.

Overall the standard of fire precautions were good and only a limited number of recommendations have been made (see Section 9).

3. **Fire risks and their control**

The principal fire risks on the premises relate to the storage and use of highly flammable paints and solvents. In recent years these have been replaced progressively by systems based on water and less flammable solvents but their use is likely to continue for the foreseeable future. Full details of control measures for highly flammable paints and solvents are contained in the separate risk assessment carried out under DSEAR.

Other fire risks are fairly normal for these types of premises. Smoking is not permitted in any part of the workshops or offices. Waste materials are stored in skips and wheelie bins in the rear compound away from the building. The compound is locked outside normal working hours.

4. **Fire detection**

Smoke detectors are situated within the corridors and staircase of the office accommodation and there are several heat detectors in the workshop. (Locations are shown on the drawings accompanying the Fire Certificate.) Any fire occurring in the workshop areas would also be likely to be detected visually. However, there is no maintenance contract for the fire detection system.

5. **Fire alarm**

Break glass alarm points are situated alongside all exit routes from the building (locations detailed in the Fire Certificate). The provision of sounders appears to be adequate and staff state that these are easily audible in all parts of the premises, including the toilets.

Records show that testing of the alarm system has only been carried out intermittently.

There is no maintenance contract for the alarm system.

The break glass point in the reception area is partially obstructed by a display stand.

6. Means of escape

Means of escape are detailed in the Fire Certificate. All exit routes are clearly indicated by the 'running man' signs. Most parts of the premises have more than one escape route, with travel distances well within the recommended limits. There are exit doors at each end of the vehicle finishing units, and the paint preparation room.

Three areas have only one escape route:

- the first floor offices, which have 14 metres travel distance to the fire protected staircase;
- the upper level of the parts department, which has a maximum 12 metres travel distance to the ground floor, from which there are two escape routes, the nearer only 3 metres away;
- the flammable liquid store which is only 3 metres long.

These are all within recommended limits.

Some items were being stored on the parts department stairway.

Emergency lighting units are situated above all of the exits from the main workshop and parts area and emergency lighting is provided in the corridors and staircase of the office.

There are no arrangements for maintenance of emergency lighting.

7. Fire evacuation arrangements

Instructions on action to be taken in the event of fire are clearly displayed throughout the premises. Recently recruited staff confirm that they have been given instruction on fire evacuation arrangements during their induction programme.

The duty receptionist is expected to call the fire service in case of fire. The fire assembly point is outside the entrance to the adjoining church and section heads are expected to conduct a roll call. The receptionist must ensure that any customers in the waiting area are escorted from the premises. However, a fire evacuation drill has not been conducted since June 2005.

8. Fire fighting equipment

An appropriate range of portable fire extinguishers is provided in the premises (as detailed in the Fire Certificate). Several foam and dry powder extinguishers are available close to the vehicle finishing units and paint preparation room and also near the solvent and paint store.

9. Recommendations

(a) Establish a contract for quarterly and annual maintenance of the fire detection and fire alarm system and also the emergency lighting.

(b) Ensure that the fire alarm is tested every week, using different call points in rotation. All tests must be recorded in the fire log book.

(c) Reposition the display stand in the reception area so that the alarm call point is not obstructed.

(d) Remove the items which are partially obstructing the stairway in the parts department. All fire escape routes must be kept clear.

(e) A fire evacuation drill must be held at least every six months, with details recorded in the fire log book. (Drills can be made more realistic by simulating the blockage of a commonly used escape route.)

Assessment conducted by: *C. D. Sparks* *I Carson*

Date: 4 October 2006

After the assessment

Review and implementation of recommendations

14.23 In some respects the process for reviewing and implementing recommendations will be similar to that for other types of assessment, as described in some detail in **CHAPTER 4: CARRYING OUT RISK ASSESSMENTS**. Many simple recommendations should be capable of being implemented quite promptly such as those relating to:

- testing of fire alarms;
- wedging open of self-closing fire doors;
- holding of fire evacuation drills;
- positioning of fire extinguishers;
- housekeeping standards;
- working practices;
- fire-related signs.

Some items may require a little longer, e.g.

- training of staff in fire fighting and other emergency action
- changes to fire evacuation arrangements.

Other recommendations may require more technical input – from the fire authority, from suppliers of specialist equipment or from both. This technical input may be required in relation to:

- the need for fire detection and alarm equipment;
- the design of such equipment;
- the adequacy of means of escape in borderline situations(especially for single escape routes);
- other fire escape or fire separation issues;
- automatic sprinklers and other fixed fire fighting systems.

Where the assessment reveals issues involving the storage and use of dangerous substances or relating to process activities, advice from the HSE may be required. (The reference section of this chapter includes some relevant publications.)

Maintaining fire precautions

14.24 *Article 17* of the Order requires that, where necessary, '. . . the premises and any facilities, equipment and devices provided under this Order . . . are subject to a suitable system of maintenance and are maintained in an efficient state, in efficient working order and in good repair.' Fire precautions can be monitored through the use of general health and safety inspection and audit programmes (as described in **CHAPTER 8: IMPLEMENTATION OF PRECAUTIONS**) or, in the case of smaller workplaces, through informal inspections.

Tests of fire equipment can usually be carried out by employees, providing they have received fairly basic training. Most employers prefer to contract out maintenance of fire equipment to specialist companies but there is no reason in law why this also cannot be done in-house, providing the staff involved are competent for the purpose. Frequencies of testing and maintenance where specific guidance is not already available (see 14.5) are suggested below.

Fire detection and alarm systems

- weekly testing of the alarm, using different alarm call points each week in rotation (many employers prefer to test at the same time each week and also to inform staff and others that a test is taking place);
- annual inspection and test by a competent person, including testing and/or inspection of smoke and heat detectors.

Emergency lighting

- testing at one to three monthly intervals;
- annual inspection and test by a competent person.

Fire evacuation procedures

- evacuation drills at intervals of 3–12 months depending upon the fire risk (drills can be made more realistic by simulating the non-availability of exit routes).

Fire fighting equipment

- annual checks by a competent person of extinguishers, sprinkler systems and other fixed fire fighting extinguishing systems.

Assessment review

14.25 Fire risk assessments, like other types of assessment, must be reviewed if they are thought to be no longer valid or if significant changes take place. Periodic reviews should be carried out in any case. Intervals of 2–5 years are recommended depending upon the level of fire risk and the potential for gradual change to take place.

Dangerous Substances and Explosive Atmospheres Regulations 2002 (DSEAR)

14.26 The *DSEAR Regulations* are concerned with protecting against risks from fire and explosion. They replaced several other regulations including the *Highly Flammable Liquids and Liquified Petroleum Gases Regulations 1972* and (apart from the dispensing of petrol from pumps) the requirement for licensing under the *Petroleum (Consolidation) Act 1928.*

Dangerous substances

14.27 The Regulations contain a detailed definition of 'dangerous substances' including:

- explosive, oxidising, extremely flammable, highly flammable or flammable substances or preparations (whether or not classified under the *CHIP Regulations*), e.g. petrol, solvents, paints, LPG;
- other substances or preparations creating risks due to their physico-chemical or chemical properties;
- potentially explosive dusts, e.g. flour, sugar, custard, pitch, wood.

Explosive atmosphere

14.28 This is defined as 'a mixture, under atmospheric conditions of air and one or more dangerous substances in the form of gases, vapours, mists or dusts, in which, after ignition has occurred, combustion spreads to the entire unburned mixture'.

Main requirements of the Regulations

14.29 The Regulations require employers to:

- Carry out a risk assessment of work activities involving dangerous substances (*Regulation 5*).
- Eliminate or reduce risks as far as is reasonably practicable (*Regulation 6*).
- Classify places where explosive atmospheres may occur into zones, and ensure that equipment and protective systems in these places meet appropriate standards, marking zones with signs, where necessary (*Regulation 7*). (This duty was phased in up to July 2006 for existing equipment and workplaces.)
- Provide equipment and procedures to deal with accidents and emergencies (*Regulation 8*).
- Provide employees with suitable and sufficient information, instruction and training (*Regulation 9*) (including details of dangerous substances present and the significant findings of the risk assessment).

Guidance on the Regulations for small and medium workplaces is available in a free HSE leaflet *Fire and explosion: How safe is your workplace?* (**REF. 8**). The HSE have published several more detailed guidance documents on the requirements of DSEAR (**REFS 9–15**). Several earlier publications are also still available to provide additional guidance if required (**REFS 16–24**).

Control measures

14.30 *Regulation 6* of the *DSEAR Regulations* requires control measures to be taken in order to reduce risks from dangerous substances. These must be implemented in a specified order of priority which can be summarised as follows:

- Reduce the quantity of dangerous substances to a minimum.
- Avoid or minimise releases of dangerous substances.
- Control releases at source.
- Prevent the formation of an explosive atmosphere (including providing appropriate ventilation).
- Collect, contain and remove any releases to a safe place or otherwise render them safe.
- Avoid the presence of ignition sources (including electrostatic discharges).
- Avoid adverse conditions that could lead to danger (e.g. by installing temperature or other controls).
- Segregation of incompatible dangerous substances.

The Regulation also requires measures to mitigate the detrimental effects of a fire or explosion (or other harmful physical effects). These include minimising the number of employees exposed and the provision of explosion pressure relief arrangements or explosion suppression equipment.

DSEAR assessment records

14.31 As yet the HSE have not provided any guidance on the format of DSEAR risk assessment records. The content should be similar to that for other types of risk assessment – details of the risks (i.e. the dangerous substances present) and a description of the control measures and other precautions in place, together with recommendations or action points resulting from the assessment.

The illustrative assessment record provided on the next page relates to the vehicle body repair business that was the subject of the fire risk assessment earlier in the chapter. The headings used in the assessment are based on DSEAR requirements.

- Dangerous substances present
- Elimination of dangerous substances

- Control measures
- Fire or explosion mitigation measures
- Places where explosive atmospheres may occur
- Accident, incident and emergency arrangements
- Information, instruction and training
- Identification of hazardous contents.

(It should be noted that this example is for illustrative purposes only – all of the requirements of DSEAR and other technical standards which are relevant to this type of establishment will not necessarily have been included.)

Sample DSEAR assessment record

14.32

BODY BEAUTIFUL REPAIR CENTRE
DSEAR RISK ASSESSMENT

Actions

1. **Dangerous substances present**

   Highly flammable solvents and flammable paints are stored and used on the premises. Safety data sheets for those currently present are kept by the workshop supervisor. 1.1

2. **Elimination of dangerous substances**

   Use of highly flammable paints and solvents has progressively decreased as they have been replaced by systems based on water and less flammable solvents. However, they are likely to remain for the foreseeable future in order to cater for older vehicles with earlier paint systems.

3. **Control measures**

   - **Storage**
     - a maximum of two 50-litre drums of highly flammable solvent and a limited number of paint containers are stored (additional materials can be obtained when required from the suppliers);
     - drums and containers must be kept firmly closed;
     - a small ramp below the door would keep any spillage inside the store;
     - the store has good natural ventilation at low and high levels; 3.1
     - lighting in the store is suitable for a Zone 1 area;
     - signs outside state 'no smoking' and 'no unprotected electrical equipment'.
   - **Paint preparation room**
     - quantities of highly flammable paints and solvents are kept to a minimum;
     - proprietary solvent dispensing container with self-closing lid provided; 3.2
     - proprietary gun wash unit with local exhaust ventilation;
     - metal waste containers with snap-shut lids provided; 3.3
     - exhaust ventilation provided for the room appears adequate (tested regularly);
     - self-closing doors with sills provide entrance to the room; 3.4
     - lighting suitable for a Zone 1 area.
   - **Vehicle finishing units**
     - local exhaust ventilation for both units appears adequate (tested regularly);
     - ventilation systems interlocked with spray guns;
     - airflow switches activate alarms if flow rate drops;
     - self-closing doors at each end of the units;
     - lighting suitable for a Zone 0 area;
     - air handling unit equipped with suitable filters and bifurcated fans;
     - ventilation discharge duct at roof level.

4. Fire or explosion mitigation measures

The storage building, paint preparation room and vehicle finishing units are all constructed of suitable fire-resisting materials (at least half-hour standard). The exhaust ducting is also half-hour fire resistant.

The roof of the storage building is lightweight, effectively providing explosion relief and the vehicle finishing units incorporate explosion relief panels in their ceilings (see the suppliers' technical literature).

5. Places where explosive atmospheres may occur

The storage building and paint preparation room are categorised as Zone 1. 5.1

The vehicle finishing units are classified as Zone 0. 5.1

Electrical equipment in these places is of the required standards.

Staff working in these areas wear anti-static overalls.

6. Accident, incident and emergency management systems

Supplies of absorbent granules, absorbent mats, rags and paper wipes are available to deal with accidental spillages of highly flammable and flammable materials.

Metal waste containers with snap-shut lids are provided and there are several metal dustbins with lids available in the rear compound. 6.1

(General fire precautions are detailed in the Fire Risk Assessment.)

7. Information, instruction and training

New staff members likely to work with highly flammable or flammable paints and solvents are briefed by the workshop supervisor about the precautions necessary in their work and the operation of the paint preparation room and vehicle finishing units (and associated equipment). They are also briefed about spillage clear-ups and made aware of where the suppliers' safety data sheets are available. 7.1

8. Identification of hazardous contents

All containers of dangerous substances are clearly identified with labels and signs, as required by the *CHIP Regulations*.

9. Recommendations (numbering relates to the sections above)

1.1 Clearly identify the safety data sheet file and place it in a prominent position in the supervisor's office.

3.1 Remove the debris blocking parts of the lower ventilation grilles of the flammable liquid store.

3.2 Only suitable self-closing solvent dispensing containers must be used in the paint preparation room – an open jug was found in use.

3.3 The metal waste containers in the paint preparation room must be emptied regularly – two were overfull, thus wedging their lids open.

3.4 The self-closing device on one of the self-closing doors to the paint preparation room required adjustment.

5.1 Provide the standard 'warning sign for places where explosive atmospheres may occur' on the doors to the flammable liquid store, paint preparation room and vehicle finishing units.

6.1 Replace the metal lids missing from two of the metal dustbins in the rear compound.

7.1 Prepare a checklist of topics to be covered during the briefing of new painters, etc. by the workshop supervisor and keep records of all staff who have received this information, instruction and training.

Assessment by: *P. Humes* *C. D. Sparks*

Date: 27 August 2003

References

14.33

1		Fire Safety: An employer's guide. HSE (1999).*
2	HSG 168	Fire safety in construction. HSE (1997)
3	BS 5446-1	Specifications for self-contained smoke alarms and point-type smoke detectors.
4	BS 5839	Fire detection and alarm systems for buildings.
5	BS 7863	Recommendations for colour coding to indicate the extinguishing media contained in portable fire extinguishers.
6	BS 5423	Portable fire extinguishers.
7	L64	Safety signs and signals. The Health and Safety (Safety Signs and Signals) Regulations 1996. Guidance on Regulations. HSE (1997)
8	INDG 370	Fire and explosion: How safe is your workplace? HSE (2002) – free leaflet.
9	L133	Unloading petrol from road tankers. (2003)
10	L134	Design of plant, equipment and workplaces. (2003)
11	L135	Storage of dangerous substances. (2003)
12	L136	Control and mitigation measures. (2003)
13	L137	Safe maintenance, repair and cleaning procedures. (2003)
14	L138	Dangerous substances and explosive atmospheres. (2003)
15	L139	Manufacture and storage of explosives. (2005)
16	HSG 140	Safe use and handling of flammable liquids. HSE (1996).
17	HSG 146	Dispensing petrol: Assessing and controlling the risk of fire and explosion at sites where petrol is stored and dispensed as a fuel. HSE (1996)
18	INDG 227	Safe working with flammable substances. HSE (1996) – free leaflet.
19	HSG 178	The spraying of flammable liquids. HSE (1998)
20	HSG 51	The storage of flammable liquids in containers. HSE (1998)
21	HSG 176	The storage of flammable liquids in tanks. HSE (1998)

22	HSG 103	Safe handling of combustible dusts: Precautions against explosions. HSE (2003)
23	CHIS 4	Use of LPG in small bulk tanks. HSE (1999)
24	CHIS 5	Small-scale use of LPG in cylinders. HSE (1999)

* This publication may have been replaced by the guides available via the websites of the Deputy Prime Minister and the Small Business Service of Business Link see **14.5** GUIDANCE ON THE ORDER.

15 Assessment of risks from asbestos in premises

In this chapter:

Introduction 15.1

The risks from asbestos 15.2

Control of Asbestos at Work Regulations 2002 15.3
Other general requirements for asbestos work 15.4
Practical guidance on asbestos work 15.5

Regulation 4 outlined 15.6
Dutyholders under Regulation 4 15.7
Possible combinations of dutyholders 15.8
Some practical implications 15.9
The duty to co-operate 15.10
Delegating tasks 15.11

Assessing the risks from asbestos 15.12
Where asbestos might be present 15.13
Survey types 15.14
Categorising materials 15.15
Which type of survey? 15.16
Aspects to be covered during assessment 15.17
Assessment review 15.18
Assessment records 15.19

The asbestos risk management plan 15.20
Factors in preparing the management plan 15.21
Risk assessment 'algorithms' 15.22
Content of asbestos risk management plans 15.23
ACM maintenance and removal 15.24
Monitoring ACMs 15.25
Providing information about ACMs 15.26
Controlling building and maintenance work 15.27
Management plan records 15.28
Managing asbestos contractors 15.29
Review, revision and implementation of the management plan 15.30

References 15.31

Introduction

15.1 The risks associated with asbestos have been recognised for many years with regulations controlling its use in manufacturing processes introduced as far back as 1931. Further regulations progressively tightened controls on work involving asbestos, with the *Control of Asbestos at Work Regulations 1987* containing a requirement for the 'assessment of work which exposes employees to asbestos'. These Regulations preceded the *COSHH Regulations* which contained many similar requirements.

The *Control of Asbestos at Work Regulations 2002* continued this requirement for a risk assessment of any asbestos work. However, *Regulation 4* of the 2002 Regulations contained a new 'duty to manage asbestos in non-domestic premises'. This requires 'dutyholders' (as defined in the Regulation) to carry out assessments 'as to whether asbestos is or is liable to be present in the premises'. Further changes are likely, as in 2005 the HSE opened a consultation process on proposals for new *Control of Asbestos Regulations* to replace the *Control of Asbestos at Work Regulations 2002*, the *Asbestos Licensing Regulations 1983* and the *Asbestos (Prohibition) Regulations 1992* with a single set of regulations.

Whilst only a limited number of employers are likely to be involved directly in work with asbestos, many more will occupy or be responsible for premises which are known or suspected to have ACMs within them. Consequently this chapter is primarily concerned with the duty to manage asbestos in premises, although it does contain some guidance and references of relevance to carrying out work with asbestos.

The risks from asbestos

15.2 Currently asbestos-related diseases cause up to 3000 deaths a year in the UK. These deaths are attributable to asbestosis (a form of pneumoconiosis which causes an irreversible scarring within the lungs), lung cancer and mesothelioma (a type of cancer affecting the lining around the lungs and sometimes the stomach). All of these diseases take many years to develop. Workers in the construction and building maintenance trades are the most affected although other activities such as shipbuilding and ship repair, work in power stations and, of course, asbestos manufacturing processes are also involved. Generally the greater the worker's exposure to asbestos then the greater the degree of risk, although it is not unknown for asbestos-related cancers to affect those who have had only limited exposure.

The risks occur when asbestos fibres become airborne and are inhaled. The *Control of Asbestos at Work Regulations* are mainly concerned with preventing fibres becoming airborne but they also contain requirements to avoid airborne fibres being inhaled – by avoiding people being in the vicinity or ensuring that people are provided with suitable respiratory protective equipment.

Fibres can become airborne for a number of reasons:

- direct work on ACMs (drilling, boring, cutting, breaking, etc.);
- disturbance or removal of ACMs;
- demolition of structures containing ACMs;

- repeated damage to ACMs (continual scrapes and knocks);
- damaged ACMs being subjected to vibration or air currents.

Control of Asbestos at Work Regulations 2002

15.3 All parts of the Regulations are now fully operative. The main changes from the previous regulations were:

- *Regulation 4* – Duty to manage asbestos in non-domestic premises (in force on 21 May 2004)

  This is outlined below and explained in some detail in the remainder of the chapter.

- *Regulation 14* – Arrangements to deal with accidents, incidents and emergencies

  Procedures, warning and communications systems must be established and information provided for such situations arising out of the use of asbestos in a work process or the removal or repair of ACM.

- *Regulation 19* – Standards for air testing
- *Regulation 20* – Standards for analysis (in force on 21 November 2004)

For both *Regulations 19* and 20 external organisations carrying out air testing or analysis work must be accredited as complying with ISO 17025 by the UK Accreditation Service. Where employers carry out this work in-house, they must make sure they apply similar standards of training, supervision and quality control to those required by ISO 17025.

Other general requirements for asbestos work

15.4 Whilst this chapter is mainly concerned with the 'duty to manage asbestos in non-domestic premises', employers should have an awareness of general requirements which must be applied to all types of work with asbestos. New duties imposed by the *Control of Asbestos at Work Regulations 2002* (the *Control of Asbestos Regulations*) are identified above. The Regulations also contain requirements for:

- identification of the type of asbestos (*Regulation 5*);
- assessment of work which exposes employees to asbestos (*Regulation 6*);
- plans of work (*Regulation 7*);
- notification of work with asbestos (*Regulation 8*);
- information, instruction and training (*Regulation 9*);

- prevention or reduction of exposure to asbestos (*Regulation 10*);
- use of control measures, etc. (*Regulation 11*);
- maintenance of control measures, etc. (*Regulation 12*);
- provision and cleaning of protective clothing (*Regulation 13*);
- duty to prevent or reduce the spread of asbestos (*Regulation 15*);
- cleanliness of premises and plant (*Regulation 16*);
- designated areas (*Regulation 17*);
- air monitoring (*Regulation 18*);
- health records and medical surveillance (*Regulation 21*);
- washing and changing facilities (*Regulation 22*);
- storage, distribution and labelling of raw asbestos and asbestos waste (*Regulation 23*);
- supply of products containing asbestos for use at work (*Regulation 24*).

The Regulations are reproduced in full together with an ACOP and related guidance in HSE booklet L27 *Work with asbestos which does not normally require a licence* (**REF. 1**). Such work, not requiring a licence, includes:

- asbestos sampling and laboratory analysis;
- a limited number of asbestos manufacturing processes;
- work on, or which disturbs, building materials containing asbestos
  - any work with asbestos cement, e.g. cleaning, painting, repair or removal
  - any work with materials of bitumen, plastic, resins or rubber;
  - 'minor work' with asbestos insulation, asbestos coating and asbestos insulating board.

'Minor work' is defined as work where either:

- any person works less than a total of 1 hour in any seven days; or
- the total time spent on the work does not exceed 2 hours.

Any more extensive work normally requires a licence from the HSE under the *Asbestos (Licensing) Regulations 1983*. Employers are permitted to carry out work at their own premises provided they use their own workers, but this option is not recommended to those not experienced in such work. All work must comply with the standards contained in an HSE ACOP *Work with asbestos insulation, asbestos coating and asbestos insulating board* (**REF. 2**). Guidance on managing contractors carrying out asbestos work is provided later in the chapter (see **15.29**).

Practical guidance on asbestos work

15.5 The HSE has published much useful guidance on asbestos work and two booklets merit particular mention for the benefit of employers who are likely to be involved in a limited range of asbestos work.

- *Introduction to asbestos essentials: Comprehensive guidance on working with asbestos in the building maintenance and allied trades* (**REF. 3**)

  This provides guidance on the principles to be followed in assessing the risks from work with asbestos and describes the range of control measures which may be necessary to carry out the work safely.

- *Asbestos essentials task manual* (**REF. 4**)

  This contains a series of task guidance sheets for common activities involving ACM which may be carried out in building maintenance and similar work, e.g. drilling holes in asbestos insulating board, removal of asbestos cement debris, removal of asbestos-containing floor tiles. The task sheets provide guidance on equipment required, appropriate PPE, preparing the work area, carrying out the task itself, cleaning arrangements, personal decontamination and clearance procedures.

Regulation 4 outlined

15.6 *Regulation 4* of the *Control of Asbestos Regulations* which imposes the duty to manage asbestos in non-domestic premises contains 11 different sub-sections. These deal with:

- *ss (1)* – The 'dutyholder';
- *ss (2)* – Co-operation with the dutyholder;
- *ss (3)*, (*4*) and (*5*) – Assessment of the presence of asbestos;
- *ss (6)* – Assessment review;
- *ss (7)* – Records of assessment (and reviews);
- *ss (8)* – Risk management plan;
- *ss (9)* – Management Plan Contents;
- *ss (10)* – Review and revision of the management plan;
- *ss (11)* – Definitions ('assessment', 'premises' and 'plan').

Domestic premises are stated in the ACOP applying to *Regulation 4* (**REF. 5**) to be private dwellings in which a person lives. Consequently, this Regulation does not apply to landlords in relation to individual houses and flats. However, they still have general duties to tenants under *HSWA 1974* and other legislation and work on ACM must still conform with the general requirements of the *Control of Asbestos Regulations*.

Non-domestic premises will include common parts of housing developments and flats such as foyers, corridors, boilerhouses, lifts and lift shafts, etc., and landlords and managing agents must apply the requirements of *Regulation 4* to such places.

In addition to the ACOP booklet *The management of asbestos in non-domestic premises* (**REF. 5**), the HSE has also published *A comprehensive guide to managing asbestos in premises* (**REF. 6**) and much of the guidance contained in this chapter is based on these booklets, together with the author's practical experience.

Dutyholders under Regulation 4

15.7 *Paragraph (1)* of *Regulation 4* of the *Control of Asbestos Regulations* defines the 'dutyholder' as:

> (a) every person who has, by virtue of a contract or tenancy, an obligation of any extent in relation to the maintenance or repair of non-domestic premises or any means of access thereto or egress therefrom; or
>
> (b) in relation to any part of non-domestic premises where there is no such contract or tenancy, every person who has, to any extent, control of that part of those non-domestic premises or any means of access thereto or egress therefrom,
>
> and where there is more than one dutyholder, the relative contribution to be made by each such person in complying with the requirements of this regulation will be determined by the nature of the maintenance and repair obligation owed by that person.

In effect, who is the dutyholder depends upon what the contract or tenancy agreement says in relation to maintenance and repair or, in the absence of such documents, who exerts practical control over such premises. As well as providing plenty of scope for lawyers, this definition could create some interesting situations, e.g. where there is local management of schools, the governing body and head teacher could be deemed the dutyholders and thus responsible for compliance with the Regulation.

Possible combinations of dutyholders

15.8 The HSE ACOP booklet (**REF. 5**) provides some useful guidance on who are likely to be the 'dutyholders' in differing sets of circumstances, dependent upon the nature of the premises and the terms of the lease. These examples are summarised below (the term 'owner' may also include an overall leaseholder).

Shared responsibilities – owner and occupier

- Owner – walls, roof, common parts;
- Employer – internal parts of premises occupied;
- Joint – management plan and responsibility to provide information.

Owner retains all responsibilities

- Owner – identify ACMs, prepare a plan, inform employers;
- Employer – pass on information where necessary.

Employer in occupation responsible

- Employer:
  - identifies ACMs, prepares a plan;
  - gives the owner (and other interested parties) access to the information.

(The *owner* becomes the dutyholder if the premises are empty.)

Responsibilities shared between many parties

- Owner, sub-lessors, employers in occupation:
  - All have a duty to co-operate (*Management Regulations, Control of Asbestos at Regulations*).
  - Best for one party to take the lead (owner, managing agent or major occupying employer).
  - Responsibility for costs likely to be shared.

Managing agent

- Owners (or leaseholders) may pass on some responsibilities to a managing agent.

  However, the owner's legal duty cannot be totally passed over.

Some practical implications

15.9 The HSE guidance refers to the sharing of the costs of compliance where one party takes the lead in complying with *Regulation 4,* in situations where responsibilities are shared between many parties. This approach should also be considered in other situations where owners and occupiers both have responsibilities. It will generally be better to take an integrated approach to the risk assessment and related survey work as well as to the subsequent risk management plan. One party (more likely the owner) may well have a greater degree of technical expertise in this area. Owners should also note that they become the dutyholder if the premises becomes empty and that they must ensure all relevant information is passed on to any new occupier (*Regulation 4(a)(c)(i)*). Whilst the occupier vacating the premises (who may have been the previous 'dutyholder') has a duty to co-operate, as detailed below, this is of little practical value to the owner if the occupiers have done a 'moonlight flit', taking any records with them.

It would seem prudent for owners, managing agents, etc. to remind tenants of their obligations under this Regulation and to attempt to gain a copy of assessment and survey records whilst the tenant is still in occupation. A contribution towards the costs of the assessment may be one means of securing this.

The duty to co-operate

15.10 *Paragraph (2)* of *Regulation 4* states that:

> Every person shall co-operate with the dutyholder so far as is necessary to enable the dutyholder to comply with his duties under this regulation.

The practical implications of this requirement are:

- Information holders must assist dutyholders (although not necessarily in meeting the costs of dutyholders' actions).
- Owners must provide relevant information to occupiers of premises.
- Architects, surveyors, building contractors must make information available to dutyholders (at a 'justifiable and reasonable cost').
- Occupiers must allow access for inspections, surveys, etc. to take place.
- Occupiers must make information available to maintenance workers, etc. (including those who have not been engaged by them).
- Occupiers must pass information on ACMs to owners on or after vacating the premises (see **15.9** above).

Again there are potential cost implications of this duty. What an architect or building contractor regards as a 'justifiable and reasonable cost' for searching through archived records 20 years old in order to check on specifications of construction materials used may not be the same as that of the person seeking the information. One potential lever in the dutyholder's favour is that the architect or contractor may well be seeking further engagements from them in the future.

In order for occupiers to provide information about ACM to maintenance workers visiting the premises they must have ready access to assessment records. This is equally important whether or not the occupier or maintenance worker is from the same organisation as the owner. The availability of appropriate records will be referred to later in the chapter.

Delegating tasks

15.11 Where dutyholders delegate work required under *Regulation 4* to others (whether to their own employees or to outside organisations), they must ensure they are capable of carrying out these tasks. Even though *the task can be delegated, the legal responsibility cannot.*

Anyone carrying out work in relation to *Regulation 4* must:

- know what they have to do;
- be able to do it safely, e.g. carry out inspections, take samples;
- have suitable competence and training for the work.

Employees and safety reps must be consulted about their appointment, either informally or through safety committee or staff meetings.

Separate organisations or persons engaged to carry out tasks must:

- have adequate training and experience for the work;
- demonstrate independence, impartiality and integrity;
- have an adequate quality management system;
- survey in accordance with MDHS 100 (**REF. 7**)

Organisations carrying out surveys must be accredited to ISO 17020; or individual surveyors have personal certification under EN 45013. Accreditation of in-house employees to carry out sampling under ISO 17020 or analysis under ISO 17025 also indicates their competence.

Assessing the risks from asbestos

15.12 *Paragraph (3)* of *Regulation 4* of the *Control of Asbestos Regulations* contains the requirement for the assessment of asbestos risks whilst *paragraphs (4)* and (5) set some parameters for the conduct of the assessment.

> (3) In order to enable him to manage the risk from asbestos in non-domestic premises, the dutyholder shall ensure that a suitable and sufficient assessment is carried out as to whether asbestos is or is liable to be present in the premises.
>
> (4) In making the assessment—
>
> (a) such steps as are reasonable in the circumstances shall be taken; and
>
> (b) the condition of any asbestos which is, or has been assumed to be, present in the premises shall be considered.
>
> (5) Without prejudice to the generality of paragraph (4), the dutyholder shall ensure that—
>
> (a) account is taken of building plans or other relevant information and of the age of the premises; and
>
> (b) an inspection is made of those parts of the premises which are reasonably accessible.

Where asbestos might be present

15.13 Asbestos was once commonly used in building materials (and in some forms of work equipment). It might be found:

- particularly in buildings from the 1950s, 1960s and early 1970s;
- in refurbishments of older buildings carried out during this period;
- possibly in more recent buildings;
- possibly in work equipment, e.g. lagging, gaskets, fire protection.

Use of asbestos progressively decreased and the importation, supply and use of ACM is now banned (apart from a few specialised applications):

- use of blue asbestos mainly ceased in about 1970;
- use of sprayed coatings decreased during the period between 1970 and 1980 and was banned since 1985;
- use of asbestos insulation board decreased after 1980 and stopped in 1985;
- use of asbestos paints and varnishes stopped in about 1988;
- asbestos-containing decorative plasters were banned in 1992;
- installation of asbestos cement was prohibited in 1999.

(However, old stock may have been used more recently than the above dates, possibly illegally, and a cautious approach should always be taken in the identification of possible ACMs.)

Potential sources of information about the possible presence (or otherwise) of ACMs include:

- drawings and material specifications;
- architects, builders, other contractors, surveyors;
- the employer's own workforce, particularly technical staff and including safety representatives (there is a legal duty to consult with employees and safety representatives under *Regulation 4*);
- records of previous surveys.

The 'Asbestos Building' depicted in the HSE publication MDHS 100 (**REF. 7**) shows the many and varied situations in which ACM might be found.

Survey types

15.14 The HSE booklet MDHS 100 *Surveying, sampling and assessment of asbestos-containing materials* (**REF. 7**) provides detailed guidance on surveys. It sets out three types of asbestos survey and contains much practical guidance on how surveys should be carried out.

Type 1 – a presumptive survey

In a Type 1 survey no sampling is carried out. There is no positive identification of ACMs and materials are only excluded if the surveyor is confident they are not ACM. All other materials are presumed to be ACM and their condition must be assessed. A Type 1 survey can be followed by a Type 2 survey at a later date. In effect the expense of sampling and analysis is deferred. However, the penalty is that management measures must be applied to materials which may eventually be found not to be ACM.

Type 2 – a sampling survey

Representative samples of suspected ACMs are collected and analysed during Type 2 surveys. MDHS 100 (**REF. 7**) contains guidance on safe sampling practices and also on what constitutes a representative sample. The condition of ACMs must also be assessed during Type 2 surveys.

Type 3 – a full access sampling and identification survey

These are normally carried out before major refurbishment or demolition work takes place. They may require access to normally inaccessible places in order to determine whether ACM might be exposed during the work. They will provide a basis for tendering for removal of ACMs and therefore the condition of ACM would not normally be assessed (apart from obvious damage or asbestos debris). A Type 3 survey and a related method statement would be a key part of a CDM Health and Safety Plan.

Categorising materials

15.15 As a result of the assessment, materials will be placed into one of three different categories:

- Those known to be ACMs
  - from specification, observation or analysis;
  - measures must be taken to manage them;
  - precautions must be taken when they are likely to be disturbed.
- Those presumed to be ACMs
  - not worth analysing now;
  - they may be inaccessible and unlikely to be disturbed;

  - measures must be taken to manage them;
  - they must be analysed (or treated as ACM) when work on them is planned or disturbance is likely.
- Asbestos-free materials
  - they are clearly something else (e.g. brick, wood, plaster);
  - confirmed as asbestos-free by very strong evidence (e.g. material specifications); or
  - confirmed asbestos-free by analysis.

As stated earlier, materials presumed to be ACM during a Type 1 survey can be categorised later during a Type 2 survey as being an ACM or asbestos-free.

Which type of survey?

15.16 Whether a Type 1 or Type 2 survey should be carried out will be determined by factors such as:

- the quality of existing information about materials used in construction;
- availability of previous survey information;
- the age of buildings (and equipment within them);
- the likely need for work to be carried out in suspect areas;
- safety considerations (e.g. difficult or dangerous access, confined spaces);
- doubts over the quality of information, e.g.
  - specified materials may not have been used;
  - previous removal of ACM may have been incomplete.

In some premises a combination of Type 1 and Type 2 surveys may be appropriate.

Where Type 1 surveys are conducted, materials must be presumed to be ACM unless there is strong evidence otherwise.

Aspects to be covered during assessment

15.17 Whichever type of survey is adopted during the assessment, the following aspects must be addressed:

- Whether it is actual or suspected ACM?
- What type of material is it?
- What condition is it in?

- Could it create airborne fibres? (E.g. due to maintenance work, damage or other disturbance.)
- Would people be exposed to these fibres?

All of these aspects will be important in developing the asbestos risk management plan.

Assessment review

15.18 Like other types of risk assessment, an assessment of risks from asbestos must be reviewed in certain circumstances. *Paragraph (6)* of *Regulation 4* of the *Control of Asbestos Regulations* states:

> The dutyholder shall ensure that the assessment is reviewed forthwith if—
>
> (a) there is reason to suspect that the assessment is no longer valid; or
>
> (b) there has been a significant change in the premises to which the assessment relates.

An assessment may be considered no longer valid because:

- the quality of previous material specifications is called into doubt;
- the quality of previous survey work is questioned;
- known or suspected ACMs are found in previously inaccessible areas.

Significant changes might relate to:

- the opening up of areas containing known or presumed ACMs;
- changes of activities which may be more likely to damage or disturb ACMs (and thus increase the risk of airborne fibres).

Assessment records

15.19 *Paragraph (7) of Regulation 4* of the *Control of Asbestos Regulations* states:

> (7) The dutyholder shall ensure that the conclusions of the assessment and every review are recorded.

The nature of assessment records will depend upon the numbers and sizes of premises involved and the scale of the presence of ACMs. Either paper-based or electronic records, or both, may be appropriate dependent on the circumstances. The records should contain detailed plans and/or sketch drawings, particularly of areas containing known or suspected ACMs. Narrative providing more detailed information will be necessary for some locations. Photographs can be particularly helpful in identifying ACM for future reference during subsequent inspections or work in affected areas. A digital camera is a particularly useful tool for feeding relevant information into computer-based records.

Consideration must be given to:

- the long-term security of records (against computer viruses, fire or accidental disposal);
- their ready availability to those who need access to information, including staff, tenants, contractors, etc. (The HSE ACOP states that a paper record or access to a computer database must be 'available on site for the entire life of the premises');
- arrangements for periodic review of the records.

Appendix 2 of the HSE booklet *Managing asbestos in premises* (**REF. 6**) contains detailed guidance on the content of records which are likely to include:

- whether materials are confirmed or presumed to be ACM;
- their location and extent;
- their degree of accessibility/vulnerability;
- the type of asbestos or material involved;
- its appearance (e.g. colour, type of finish);
- the condition of the material;
- details of any areas not accessed during the assessment.

The asbestos risk management plan

15.20 *Paragraphs (8)* and (9) of *Regulation* 4 of the *Control of Asbestos Regulations* require dutyholders to prepare an asbestos risk management plan.

> (8) Where the assessment shows that asbestos is or is liable to be present in any part of the premises the dutyholder shall ensure that—
>
> (a) a determination of the risk from that asbestos is made;
>
> (b) a written plan identifying those parts of the premises concerned is prepared; and
>
> (c) the measures which are to be taken for managing the risk are specified in the written plan.

(9) The measures to be specified in the plan for managing the risk shall include adequate measures for—

(a) monitoring the condition of any asbestos or any substance containing or suspected of containing asbestos;

(b) ensuring any asbestos or any such substance is properly maintained or where necessary safely removed; and

(c) ensuring that information about the location and condition of any asbestos or any such substance is—

(i) provided to every person liable to disturb it, and

(ii) made available to the emergency services.

Factors in preparing the management plan

15.21 In developing the asbestos risk management plan, account must be taken of:

- What is the risk of asbestos fibres becoming airborne?

  Are ACMs

  - already damaged or deteriorating?
  - likely to be disturbed? (E.g. during regular repair or maintenance.)
  - likely to be disturbed or damaged in normal use? (E.g. by fork lift trucks, vehicles, trolleys.)
  - susceptible to vandalism?

- What is the human exposure potential?

  - how frequently is the area used?
  - what is the average time it is used?
  - how many people would be exposed?

Risk assessment 'algorithms'

15.22 The HSE booklet *Managing asbestos in premises* (**REF. 6**) contains algorithms to assist in determining priorities by quantifying the level of risk. These take account of:

- Material assessment

  The possibility of fibres becoming airborne, considering factors such as:

  - the product type (e.g. vinyl floor tiles, sprayed lagging, etc.);
  - the extent of any damage or deterioration of the material;

  - its surface treatment (e.g. composite, enclosed, unsealed);
  - the asbestos type (e.g. chrysotile, crocidolite, amosite).
- Priority assessment

  Factors to be considered are

  - normal occupant activities in the area (and scope for them disturbing the ACM);
  - the likelihood of disturbance of the ACM (the position and size of ACM areas);
  - human exposure potential (occupancy, frequency and times of use of the location);
  - maintenance activities which may disturb the ACM.

These are explained in Appendices 2 and 3 of the booklet and several worked examples are provided in Appendix 4.

Whilst some may find the use of these algorithms helpful, it is not a requirement of the Regulations that they be used. The author has the same scepticism about them as expressed earlier in relation to risk rating matrices (see **CHAPTER 7: SPECIALISED RISK ASSESSMENT TECHNIQUES**). They are certainly not essential in drawing conclusions from the risk assessment process and preparing an adequate risk management plan.

Content of asbestos risk management plans

15.23 The plan must include adequate measures for:

- ensuring ACMs are properly maintained or, where necessary, safely removed;
- monitoring the condition of known or suspected ACMs;
- ensuring information about the location and condition of ACMs is
  - provided to every person liable to disturb it;
  - made available to the emergency services.

ACM maintenance and removal

15.24 Where known or presumed ACMs are in relatively good condition and are unlikely to be disturbed, there is no need to remove them. However, consideration must be given to various maintenance options, so that any risk from ACMs is adequately controlled. These might involve:

- labelling or colour coding ACM areas
  - to prevent disturbance by persons unaware of their nature.

- protecting or enclosing vulnerable areas of material from mechanical damage
  - using protective barriers or bollards;
  - by airtight enclosures.
- sealing or encapsulating the ACM using
  - surface coatings;
  - penetrating encapsulants;

  (fire resistance may be a consideration, however).
- repairing slight damage to small areas
  - by patching, sealing or encapsulation.

Removal may be the best option if the ACM is:

- not in good condition;
- vulnerable to damage or vandalism;
- likely to be involved in major building work (or demolition).

Removal must normally be by a licensed contractor (although this is not a legal requirement for asbestos cement or bonded materials – see **15.4**). Guidance on managing asbestos contractors is provided in **15.29**.

Appendix 5 of *Managing asbestos in premises* (**REF. 6**) provides worked examples of how consideration of the various maintenance options or possible removal can be carried out.

Monitoring ACMs

15.25 Where ACMs are left in place, their condition must be monitored regularly and the results recorded. The frequency should reflect the type and condition of the ACM and the potential for it to be disturbed or damaged. Guidance in the HSE ACOP booklet (**REF. 5**) states: 'As a minimum, the material should be checked every six to twelve months even if it is in good condition and not going to be disturbed . . .'.

However, *Managing asbestos in premises* (**REF. 6**) provides slightly different guidance: 'The time period between monitoring . . . would not be expected to be more than 12 months in most cases. ACMs in remote locations, with little or no activity, can be inspected infrequently. For example, an asbestos insulating board ceiling in a remote unoccupied building may only need inspecting once every 12 months *or even less frequently if the building is rarely entered*.'

Monitoring is likely to involve a visual inspection for signs of disturbance, scratches or cracking, broken edges, cracked or peeling surface treatment and nearby debris. Comparisons with photographs taken earlier can be very useful in monitoring long-term deterioration. Where necessary, appropriate remedial action must be initiated – this might involve some of the maintenance options referred to in **15.24** above, or even possible removal.

Providing information about ACMs

15.26 It is essential that adequate information about the known or possible presence of ACMs is provided to those who may be at risk. Such persons will include:

- All staff

  Staff should be provided with information about

  - the nature and extent of asbestos-related risks (from airborne fibres);
  - how to avoid risks (leave ACM in good condition well alone);
  - where ACMs are known or presumed to be present;
  - how to report concerns about possible ACMs (or about work involving ACMs).

- Maintenance and building workers

  Maintenance employees and regular contractors carrying out maintenance or building work should be given information as above and also

  - further details about known and presumed ACMs;
  - reminded of the need to check before working on any suspected ACM;
  - where further information about ACM is available;
  - safe methods for working with ACMs;
  - activities prohibited on known or presumed ACMs, e.g. sanding or grinding.

  Detailed guidance on safe working methods is available in two HSE booklets: *Asbestos essentials task manual: Task guidance sheets for the building maintenance and allied trades* (**REF. 4**) and *Introduction to asbestos essentials: Comprehensive guidance on working with asbestos in the building maintenance and allied trades* (**REF. 3**).

- Emergency services

  The fire service is most likely to be affected. They should be approached to see what information they want.

Controlling building and maintenance work

15.27 Exerting effective control over building and maintenance work in premises where ACMs may be present is an important part of the asbestos risk management plan. Indeed, such control should be exerted at an early stage – even before a detailed assessment of risks from asbestos has been carried out. This can be particularly important where an employer is responsible for a number of premises, some of which may not have anyone with detailed knowledge of asbestos risks or building and maintenance work available locally. Appropriate control measures may involve:

- One person controlling maintenance or building work (whether in-house or by contractors).
- Limiting the number of contractors doing maintenance or building work.
- Ensuring ACM information is readily available at all relevant locations (see **15.19**).
- Use of locked doors and/or related signs, e.g. 'Asbestos risk. No access without permission. Contact xxxx on extension 1234.'
- Controlling activities through permit to work systems.

Management plan records

15.28 The asbestos risk management plan must be recorded and should contain:

- details of the premises (or parts) it is concerned with;
- where information about known or presumed ACMs is available;
- arrangements for monitoring known or suspected ACMs;
- progress of maintenance and/or removal programmes;
- how information about ACMs is to be provided to those who may be at risk;
- how possible exposure to ACMs is controlled, e.g. locked doors and signs, permits to work;
- responsibilities for implementation of the plan;
- how the plan is to be reviewed;

The record of the assessment of risks from asbestos can be integrated with the asbestos risk management plan into a single document.

Managing asbestos contractors

15.29 Appendix 6 of *Managing asbestos in premises* (**REF. 6**) provides detailed guidance on this subject, including a checklist to assist in defining the required contents of method statements. Key elements in the management process are:

- Identify potential licensed contractors (from the HSE list, trade associations, direct recommendations).

- Provide the potential contractors with information about the work required.
- Ask them for relevant information
  - a copy of their licence;
  - details of their insurances;
  - training records for employees to be involved in the work;
  - client references from recent work they have carried out;
  - details of any HSE or local authority enforcement action;
  - a method statement and any other specific information about the job.
- Review the responses and select a contractor accordingly.
- Meet with the contractor and discuss their method statement and plan of work.
- Monitor that the work is carried out to satisfactory standards.
- Ensure a certificate of reoccupation is provided by an accredited laboratory.

Review, revision and implementation of the management plan

15.30 *Paragraph (10)* of *Regulation 4* of the *Control of Asbestos Regulations* contains fairly standard requirements for the plan to be reviewed and revised and also requirements for its implementation.

> The dutyholder shall ensure that –
>
> (a) The plan is reviewed and revised at regular intervals, and forthwith if—
>
> (i) there is reason to suspect that the plan is no longer valid, or
>
> (ii) there has been a significant change in the premises to which the plan relates;
>
> (b) the measures specified in the plan are implemented; and
>
> (c) the measures taken to implement the plan are recorded.

It may be concluded that the plan is no longer valid because its measures are considered to be ineffective or unnecessarily onerous, in the light of experience. For example, inspection frequencies may need to be increased or decreased. Changes justifying a review may involve changes in work activities or in the locations where they are carried out. The HSE guidance in the ACOP (**REF. 5**) recommends a general review every six months, even if there have been no changes. However, the nature and extent of a review will depend upon the size and complexity of the premises and the quantities of ACM involved.

References

(All HSE publications)

15.31

1	L27	Work with asbestos which does not normally require a licence. ACOP and Guidance (2002).
2	L28	Work with asbestos insulation, asbestos coating and asbestos insulation board. ACOP and Guidance (2002)
3	HSG 213	Introduction to asbestos essentials: Comprehensive guidance on working with asbestos in the building maintenance and allied trades (2001)
4	HSG 210	Asbestos essentials task manual: Task guidance sheets for the building maintenance and allied trades (2001)
5	L127	The management of asbestos in non-domestic premises. ACOP and Guidance (2002)
6	HSG 227	A comprehensive guide to managing asbestos in premises (2002)
7	MDHS 100	Surveying, sampling and assessment of asbestos-containing materials (2001)

16 Assessment of work at height

In this chapter:

Introduction 16.1

The Regulations summarised 16.2

Avoidance of risks from work at height 16.3

Work equipment for work at height 16.4

Selection of work equipment for work at height 16.5

Requirements for particular work equipment 16.6

Inspection of work equipment 16.7

Use of ladders 16.8

Planning and preparation 16.9

Making the assessment 16.10

Recording the assessment 16.11

Construction projects 16.12

Permits to work 16.13

Dynamic risk assessments 16.14

References 16.15

Introduction

16.1 The *Work at Height Regulations 2005* came into force on 6 April 2005. They replaced other requirements dealing with work at height, including parts of the *Construction (Health, Safety and Welfare) Regulations 1996* and the *Workplace (Health, Safety and Welfare) Regulations 1992*. In introducing the regulations, the HSE referred to the annual accident toll in the UK from falls from height – between 50 and 60 fatalities and around 4000 major injuries.

The Regulations themselves do not contain a separate requirement for risk assessment. However, particularly in *Regulation 6* (Avoidance of risks from work at height) and *Regulation* 7 (Selection of work equipment for work at height), the Regulations introduce specific criteria which must be taken into account when carrying out a risk assessment under *Regulation 3* of the *Management Regulations*. Several other Regulations also contain specific requirements of relevance during a risk assessment. The hierarchical approach to avoidance of risk contained in

Regulation 6 matches that taken in other regulations, particularly the *COSHH Regulations* (see 9.11: HIERARCHY OF MEASURES).

The Regulations summarised

16.2 Various terms are defined in *Regulation 2* including:

construction work – this has the meaning assigned to it in the CHSW Regulations 1996 and includes many types of work not traditionally regarded as construction, e.g. electrical installation;

fragile surface – a surface liable to fail if any reasonably foreseeable loading were to be applied to it;

ladder – includes a fixed ladder and a stepladder;

work at height –

- work in any place, including . . . at or below ground level;
- obtaining access to or egress from such place while at work, except by a staircase in a permanent workplace,

where if measures required by these Regulations were not taken, a person could fall a distance liable to cause personal injury.

This last definition replaces the 2 metres fall height contained in previous definitions – potential falls from any height must be considered.

Under *Regulation 3* requirements imposed by the Regulations on an employer apply in relation to work:

(a) by an employee of his; or

(b) by any other person under his control, to the extent of his control.

Similar requirements are imposed on the self-employed. There are exclusions from the regulations in respect of the ship-board activities of a ship's crew, dock operations, fish loading processes and caving or climbing.

Organisation and planning (*Regulation 4*) – Employers must ensure work at height is properly planned, appropriately supervised and carried out in a manner which is, so far as is reasonably practicable, safe. Planning must include the selection of work equipment in accordance with *Regulation* 7 and planning for emergencies and rescue. Work at height must be carried out only when weather conditions do not jeopardise health and safety, although these weather restrictions do not apply where members of the police, fire, ambulance or other emergency services are acting in an emergency.

Competence (*Regulation 5*) – Employers must 'ensure that no person engages in any activity, including organisation, planning and supervision, in relation to work at height or work equipment for use in such work unless he is competent to do so or if being trained, is being supervised by a competent person'.

Avoidance of risks from work at height (*Regulation 6*) – The hierarchical approach required by this regulation is explained fully in 16.3 below.

Selection of work equipment for work at height (*Regulation* 7)

Requirements for particular work equipment (*Regulation 8*) – The requirements of both of these regulations are set out in 16.5 and 16.6 below.

Fragile surfaces (*Regulation 9*) – Employers must ensure that no person at work passes across or near, or works on, from or near, a fragile surface unless it is reasonably practicable to carry out work safely and under appropriate ergonomic conditions. If this is not reasonably practicable, suitable and sufficient platforms, coverings, guard rails or similar means of support must be provided and used so far as is reasonably practicable, and where a risk still remains, suitable and sufficient measures must be taken to minimise the distances and consequences of a fall.

Prominent warning notices must be affixed at the approach to fragile surfaces so far as is reasonably practicable; or persons made aware by other means, although this does not apply where members of the police, fire, ambulance or other emergency services are acting in an emergency.

Falling objects (*Regulation 10*) – Employers must, where necessary to prevent injury, take suitable and sufficient steps to prevent, so far as is reasonably practicable, the fall of any material object. Where this is not reasonably practicable (or is otherwise necessary), suitable and sufficient steps must be taken to prevent any person being struck. Materials or objects must not be thrown or tipped from height in circumstances where it is liable to cause injury, and materials and objects must be stored to prevent risks arising from their collapse, overturning or unintended movement.

Danger areas (*Regulation 11*) – Where workplaces contain areas where there is a risk of persons at work falling a distance or being struck by a falling object, liable to cause personal injury, the workplace must, so far as is reasonably practicable, be equipped with devices preventing unauthorised persons from entering the area. Danger areas must be clearly indicated. (This regulation creates a clear requirement for barriers and signs to be in place where appropriate.)

Inspection of work equipment (*Regulation 12*) – This requirement is explained more fully in 16.7 below.

Inspection of places of work at height (*Regulation 13*) – Every employer must, so far as is reasonably practicable, ensure that the surface and every parapet, permanent rail or other such fall protection measure of every place of work at height are checked on each occasion before the place is used.

Duties of persons at work (*Regulation 14*) – Persons at work must report to those under whose control they are working any activity or defect, relating to work at height, which they know is likely to endanger the safety of themselves or anyone else. They must also use work equipment or safety devices provided for work at height in accordance with training they have received and any instructions provided.

Regulations 15–19 – deal with exemptions, amendments and revocation of other regulations, etc.

Avoidance of risks from work at height

16.3 *Regulation 6* of the *Work at Height Regulations 2005* sets out a hierarchy of measures which must be considered when carrying out a risk assessment in respect of work at height. First preference must be given to the first measure of the hierarchy, only moving down the list as each measure proves not to be reasonably practicable.

The hierarchy contained in *Regulation 6* can be summarised as follows:

1. **Avoiding carrying out the work at height**
   e.g. pruning high hedges or cleaning high windows using long-handled tools, assembling the components of timber-framed roofs at ground level
2. **Using existing places of work or means of access or egress**
   e.g. fixed platforms or stairways
3. **Using sufficient work equipment to prevent a fall occurring**
   e.g. scaffolding, tower scaffolds, mobile elevating work platforms
4. **Minimising the distance and/or consequences of a fall**
   e.g. by use of suitable fall arrest equipment such as harnesses, safety nets, foam bags or airbags.

Existing places of work and means of access or egress must comply with *Schedule 1* to the Regulations and both fall prevention equipment and fall arrest equipment must conform to standards set out in other Schedules (see 16.6 below).

The Regulation also requires that additional training and instruction, and any other additional measures must be provided, so far as is reasonably practicable, to prevent falls liable to cause personal injury.

Some work at height involves the risk of falls through fragile surfaces which are dealt with by *Regulation 9* (summarised in 16.2 above). However, in effect, *Regulation 9* requires a similar hierarchical approach to avoiding risks as that contained in *Regulation 6*.

Work equipment for work at height

16.4 Whether or not a particular type of work equipment is going to be reasonably practicable to use for a specific type of work activity is a vital part of the risk assessment process. However, the Regulations also contain several other requirements for work equipment for work at height which must be considered during the risk assessment. The most important of these are:

- selection of work equipment for work at height (*Regulation 7*);
- requirements for particular work equipment (*Regulation 8*);
- inspections of work equipment (*Regulation 12*).

Selection of work equipment for work at height

16.5 In selecting work equipment for work at height, collective protection measures must be given priority over personal protection. Consequently, whilst *Regulation 6* would require guard rails to be used in preference to fall arrest equipment, *Regulation 7* would require use of a safety net to be given priority over an individual fall arrest harness.

The Regulation also sets out various criteria which must be taken account of in selecting work equipment for use at height. These are:

- working conditions and risks at the place where the equipment is used;
- the distance to be negotiated (for access and egress);
- distance and consequences of a fall;
- duration and frequency of use;
- need for easy and timely emergency evacuation and rescue;
- additional risks posed by the use, installation or removal, etc. of the work equipment (or by evacuation and rescue from it);
- other provisions of the regulations.

Equipment for work at height must have characteristics (including dimensions) appropriate for the work and foreseeable loadings and allow passage without risk, as well as being the most suitable work equipment, having regard to the requirements of *Regulation 6*.

Some of the above criteria are often overlooked. Whilst scaffolding or a mobile elevating work platform are undoubtedly safer than a ladder for carrying out work at height, there may not be sufficient space to use them and it will take much longer to get scaffolding into place. Erecting scaffolding involves manual handling risks and possible work at height, whilst rough or sloping terrain around the work location creates risks for the mobile platform. For short term, low risk work at relatively low levels, a ladder may well prove the safest option (see 16.8 below).

Requirements for particular work equipment

16.6 The Regulations contain several Schedules each with a variety of specific requirements which must be complied with in respect of particular types of work equipment:

- Guard rails, toe boards, barriers, etc. – *Schedule 2.*
- All work platforms – *Schedule 3 Part 1.*
- Scaffolding – *Schedule 3 Part 2.*
- Nets, airbags, other collective fall arrest equipment – *Schedule 4.*
- Personal fall protection systems – *Schedule 5 Part 1.*
- Work positioning systems – *Schedule 5 Part 2.*
- Rope access and positioning techniques – *Schedule 5 Part 3.*
- Fall arrest systems – *Schedule 5 Part 4.*
- Work restraint systems – *Schedule 5 Part 5.*
- Ladders – *Schedule 6.*

Further detail on the contents of these Schedules is available in the Regulations themselves, whilst the References section of this chapter identifies several HSE publications of relevance. Guidance on factors to be taken into account in assessing the suitability of ladders for work at height is contained in 16.8 below.

Inspection of work equipment

16.7 The requirements of *Regulation 12* only apply in respect of work equipment covered by *Regulation 8* and *Schedules 2–6*. However, *Regulation 13* also contains a requirement for parapets, permanent guard rails, etc. to be checked before places of work at height are used. Several possible situations requiring inspection are covered by the Regulation, all of which must be taken into account during the risk assessment process.

Installation or assembly

Where the safety of work equipment depends on how it is installed or assembled, it must be inspected prior to being used in any new position. This requirement would be particularly important in respect of fall arrest equipment such as safety nets, foam bags or airbags.

Conditions causing deterioration

Equipment exposed to conditions causing deterioration which is liable to result in dangerous situations must be inspected at suitable intervals. This could apply to permanent or portable ladders left exposed to the weather (or the effects of chemicals), or to portable ladders subject to damage through transportation and use. The frequency of inspections should relate to the speed at which deterioration may occur.

Exceptional circumstances

Work equipment subject to exceptional circumstances liable to jeopardise safety must also be inspected. Such circumstances might involve impact (e.g. from a vehicle or a suspended load), the equipment being dropped from a height, accidental overloading of the equipment (e.g. by materials or equipment) or exceptional weather (e.g. strong winds or heavy snowfall).

Construction working platforms

Working platforms used for construction work and from which a person could fall 2 metres or more must not be used in any position unless inspected in that position within the previous 7 days. In the case of a mobile working platform, inspection on that site is acceptable. *Schedule* 7 to the *Work at Height Regulations* specifies the particulars which must be contained in the inspection report. (These inspection and reporting requirements continue those which were previously in the *CHSW Regulations*.)

Forms of inspection

In this Regulation 'inspection means such visual or more rigorous inspection by a competent person as is appropriate for safety purposes', and 'includes any testing appropriate for those purposes'. Clearly a different type of inspection is likely to be appropriate for a portable ladder as opposed to a safety net and the competences required for the inspection will differ accordingly. Regular testing of the material of safety nets is necessary to ensure that it has not deteriorated due to the effects of weather or use.

Inspection reports

The result of each inspection must be recorded and kept until the next inspection is recorded. Reports in respect of construction working platforms must be kept at the site where the inspection was carried out until the construction work is completed and thereafter at the employer's office for 3 months. Employers must ensure that no work equipment leaves their undertaking or, if obtained from the undertaking of another person, is used in their undertaking, unless accompanied by physical evidence that the last inspection has been carried out.

Use of ladders

16.8 Both prior to and since the introduction of the *Work at Height Regulations* there have been incorrect claims that the Regulations banned the use of ladders, forcing the HSE to publicly refute such claims on several occasions. However, before using a ladder (the term also includes stepladders), consideration must be given to the hierarchical approach contained in *Regulation 6* and the criteria for selection of work equipment set out in *Regulation 7*.

The HSE leaflet 'Safe use of ladders and stepladders. An employers' guide' (**REF. 2**) goes into more detail on factors to be taken into account in determining whether an activity is suitable to carry out from a ladder. Such factors include:

- use in one position for a maximum of 30 minutes;
- use for light work only (carrying more than 10 kg would need to be justified by a more detailed risk assessment);
- the availability of a handhold on the ladder or stepladder;
- being able to maintain three points of contact (hands and feet) at the working position, other than for brief periods of time;
- the ladder not being overloaded;
- the activity not involving overreaching (the user's belt buckle should be between the stiles of the ladder and both feet on the same rung);
- avoiding side loading on stepladders (e.g. side on drilling through solid materials).

Other factors to be considered should include:

- the potential height of a fall (and the surface which would be fallen onto);
- the suitability of the ground or floor surface, and the vertical support surface;
- how well the ladder can be secured in position;
- risks of vehicles or pedestrians colliding with the ladder.

Whilst tying ladders to a suitable point is preferable, current HSE thinking places both the use of an effective stability device and wedging the ladder ahead of having someone footing the ladder.

All of the above considerations influence whether the activities should be carried out from a ladder. There are also many other factors governing the safe use of ladders, for example:

- ensuring that stepladders are fully extended;
- using ladders at the correct angle (1 unit out for every 4 up);
- the suitability of weather conditions;
- pre-use inspections of ladders;
- training of staff in inspection and safe use.

For many types of situations staff will need to be trained to make their own dynamic risk assessments of whether or not it is safe to use a ladder (see 16.14 below and **17.8: DYNAMIC RISK ASSESSMENTS**).

Planning and preparation

16.9 Much of the practical guidance given in **CHAPTER 4: CARRYING OUT RISK ASSESSMENTS** is of relevance to risk assessment of work at height. Differing types of work at height need to be included in the assessment, such as:

- production and other core activities;
- installation, maintenance, repair and testing of equipment and services;
- cleaning and redecoration;
- access to stored equipment, materials and records;
- access onto vehicles and containers;
- inspection activities.

As far as possible, these should be identified prior to the assessment commencing. It will also be necessary to identify what types of equipment are available for use in work at height, e.g.

- ladders and stepladders
- tower scaffolds
- mobile elevating work platforms
- harnesses and other fall arrest equipment.

In addition to deciding which (if any) of these types of equipment are most suitable for the various types of work at height, the risk assessment must ensure that the equipment and its use meet the standards specified in the Schedules to the Regulations and that appropriate inspection arrangements are in place.

Relevant guidance material from the HSE and other sources should be gathered together. Several HSE publications are listed in the References section of this chapter and the HSE website contains additional useful information.

Making the assessment

16.10 As for other types of assessment, visits to the workplace will be essential to observe work at height being carried out or see where it takes place. Some work at height will involve work outdoors and, for some employers, much may take place away from their own premises.

Discussions will need to take place with those who carry out work at height, those who supervise and manage them, and safety representatives. These discussions should centre around aspects such as:

- whether the activities and conditions observed are representative of the norm;
- the possible variations in work activities, environments and conditions;
- practical aspects associated with using different types of work equipment;
- additional types of equipment which might be safer or more practical;
- awareness of staff of risks associated with work at height and related precautions (including formal training received);
- staff capabilities of conducting dynamic risk assessments and carrying out inspections of equipment;
- concerns of staff about certain types of work at height.

Recording the assessment

16.11 Risk assessments of work at height can be recorded in many different formats including all those illustrated in **CHAPTER 5: ASSESSMENT RECORDS**. Whilst the standard formats used in **5.4: ILLUSTRATIVE ASSESSMENTS** or standard operating procedures (see **5.8: STANDARD OPERATING PROCEDURES**) may be appropriate for routine activities, either employees' handbooks or contractors' manuals (see **5.6: EMPLOYEES' GUIDES OR HANDBOOKS** and **5.7: CONTRACTORS' MANUALS**) may be more suitable where there is more variation involved.

A sample of a risk assessment record is provided later in this chapter. It should be noted that this sample deals principally with the selection of equipment for access to carry out work at height. The full assessment should contain details of additional equipment features (e.g. safety devices for ladders), training requirements and standards for using each of the different types of equipment.

Construction projects

16.12 Many construction activities involve work at height and it is often necessary to carry out a risk assessment in advance of a construction project taking place. Such projects will often be subject to the *CDM Regulations* and it will be necessary for a CDM Health and Safety Plan to be prepared (see **17.16: CDM HEALTH AND SAFETY PLANS**). Such a plan will need to take account of the requirements of the *Work at Height Regulations*:

- activities involving work at height;
- equipment required to avoid risks of falls from height;
- working methods to avoid or reduce risks;

- planning of the work to reduce risks or ensure the availability of necessary equipment;
- work being carried out by sub-contractors.

Sub-contractors will normally be required to submit their own risk assessments or method statements in respect of any work at height. These may well be of a generic type and there will often be a need to adapt or extend them to take account of the circumstances of an individual project. The principal contractor will also need to ensure that the requirements of sub-contractors are taken into account in the planning of the project as a whole, e.g. by leaving scaffolding in place or ensuring the availability of mobile elevating work platforms for use by sub-contractors installing security equipment or other finishing trades. The sample risk assessment at the end of this chapter refers to the use of a 'Pre-contract Health and Safety Checklist'. This is another way in which a risk assessment can be recorded prior to an installation project commencing.

DUNCAN SMITH ALARMS

RISK ASSESSMENT

PAGE 1 OF 4

NO. 2 WORK AT HEIGHT

CROSS REFERENCES RISK ASSESSMENTS 3 (USE OF LIFTING EQUIPMENT); 4 (MANUAL HANDLING); 6 (SITE-RELATED RISKS)

DETAIL OF RISKS	EXPECTED PRECAUTIONS
The installation, maintenance and repair of equipment involves work at height with its attendant risks. Installation activities mainly involve (in descending order of risk): – external installation of CCTV cameras – external installation of bell boxes – internal installation work (e.g. detectors, cable running, etc.) Adjustments, inspections and minor repairs normally involve a lesser degree of risk because of the lower amounts of manual handling and physical effort involved.	The Company 'Pre-contract Health and Safety Checklist' is completed prior to every new installation project commencing and a review of the need for work at height will take place at this stage. Some high CCTV cameras are mounted on towers which can be winched down. On 'new build' projects, arrangements will be made with the main contractor to use scaffolding or mobile elevating work platforms which are already available, where this is reasonably practicable. In other cases the Company has set the following limits for working from ladders: <u>External work</u> (from ladders) • installation of CCTV cameras and other heavier items – feet at a maximum height of 3 metres (i.e. working height approx. 4.5 metres) • installation of bell boxes and similar equipment; carrying out adjustments, inspections and minor repairs – feet at a maximum height of 4 metres (i.e. working height approx. 5.5 metres) <u>Internal work</u> (from stepladders) – feet at a maximum height of 2.5 metres (i.e. working height approx. 4 metres) For work above these heights or where ladders are unsuitable, either scaffolding (inc. tower scaffolds) or mobile elevating work platforms must be used. Details of safety precautions (inc. training) required in the use of the above equipment are contained in the remainder of this risk assessment.

Permits to work

16.13 In industrial environments, it will not be possible to carry out a detailed risk assessment for every possible type of work at height which may need to be carried out in the future. Sometimes work at height may need to be considered together with other significant risks, e.g. electrical work on an overhead travelling crane in a chemical plant. It is common to control such types of work with a 'Permit to Work' – as described in 17.10–17.15.

Whilst the legal requirements and criteria to be considered are the same as for other assessments of work at height, the risk assessment is carried out by the permit issuer and recorded on the permit form.

Dynamic risk assessments

16.14 In installation and maintenance activities, staff will often be working away from their base and have to assess risks from work at height and the suitability of available equipment to carry out that work. This process which is often known as dynamic risk assessment is described more fully in 17.8.

However, where staff are expected to carry out dynamic risk assessments it is essential that they have received appropriate training so that they can:

- recognise risks involved in work at height (relating to their work activities and environments);
- identify which types of equipment are most appropriate for which circumstances;
- use that equipment safely;
- make appropriate judgements in changing circumstances e.g. adverse weather.

The availability of relevant reference material, e.g. handbooks, manuals, checklists will also be important in helping staff carry out a dynamic assessment effectively.

References

(HSE publications)

16.15

1	INDG 401	Work at Height Regulations 2005. An employers' guide (2005)
2	INDG 402	Safe use of ladders and stepladders. An employers' guide (2005)

3	CIS 10	Tower scaffolds (2005)
4	HSG 150	Health and safety in construction (1996)
5	HSG 33	Health and safety in roof work (1998)
6	CIS 49	General access scaffolds and ladders (2003)
7	AIS 32	Preventing falls from fragile roofs in agriculture (1999)
8	DIS 7	Safe working on top of containers on board ship (2000)
9	ETIS 6	Working at heights in the broadcasting and entertainment industries (1998)
10	MISC 611	Safety in window cleaning using suspended and powered access equipment (2003)
11	MISC 612	Safety in window cleaning using rope access techniques (2003)
12	MISC 613	Safety in window cleaning using portable ladders (2003)

Updating of some of the above and additional guidance is expected following the *Work at Height Regulations 2005*.

17 Risk assessment related concepts

In this chapter:

Introduction 17.1

Safe systems of work 17.2
What is a safe system of work? 17.2
Task safety instructions 17.3
Task procedures 17.4
Factors in establishing a safe system of work 17.5
Development of task procedures 17.6
Implementation of safe systems of work 17.7

Dynamic risk assessments 17.8
The 'safe person' 17.9

Permits to work 17.10
Designation of permit situations 17.11
The permit sequence 17.12
Permit to work forms 17.13
The permit issuer 17.14
Authorisation of permit issuers 17.15

CDM health and safety plans 17.16

Method statements 17.17

References 17.18

Introduction

17.1 Risk assessment techniques are at the heart of the effective management of health and safety. This chapter examines several important concepts which involve risk assessment. These are:

- Safe systems of work
- Dynamic risk assessments
- Permits to work
- CDM Health and Safety plans
- Method statements.

Safe systems of work

What is a safe system of work?

17.2 Under *section 2(2)(a)* of *HSWA 1974* employers have a duty to ensure

> the provision and maintenance of plant and systems of work that are, so far as is reasonably practicable, safe and without risks to health.

Safe systems of work are also required directly or indirectly under several codes of Regulations including those applying to asbestos, carcinogens (COSHH), confined spaces, electricity and lifting operations. Employers also have duties under common law to establish safe systems of work.

A safe system of work can be defined as:

> the work method resulting from an assessment of the risks associated with a task and the identification of the precautions necessary to carry out the task in a safe and healthy way.

The risk assessment may result in the elimination of risks or in the identification and application of controls at source, e.g. guarding, enclosure, the use of local exhaust ventilation. However, a safe system of work will be necessary to ensure that those controls are properly applied and that any residual risks are adequately controlled.

The degree of formality necessary in identifying and defining a safe system of work will depend on factors such as:

- the level of risk involved;
- the frequency of the task;
- its complexity and variability;
- the capabilities of those performing the task;
- the complexity of the precautions required.

There are several ways in which a safe system of work may be defined including *task safety instructions* and *task procedures* (referred to below) together with *permits to work* and *method statements* (covered later in the chapter). The word 'task' is preferred to 'job' as the latter can be confused with 'occupation'.

In some situations only a relatively short time might be available to establish a safe system of work, requiring a *dynamic risk assessment* to be carried out in the workplace (see **17.8: DYNAMIC RISK ASSESSMENTS**). In others, a more formalised risk assessment will be appropriate, applying the types of approach already described elsewhere in the book.

Task safety instructions

17.3 A task safety instruction is a means of conveying essential safety information to those carrying out a task. It can use a variety of forms of communication:

- informal oral instructions, e.g.
  - make sure someone is footing that ladder;
  - some of the castings are hot, you'll need to wear gloves.
- Safety signs or notices, e.g.
  - wear eye protection when operating this machine;
  - fragile roof – crawling boards must be used.
- Written safety rules or key points
  - summarising the risks involved and the precautions required.

Task procedures

17.4 A task procedure provides a step-by-step description of how a task should be performed. As well as including operational instructions and health and safety requirements, it can also contain efficiency and quality requirements relating to the task. Such formal procedures are more commonly prepared for routine production tasks but they may also be appropriate for some maintenance activities and emergency or breakdown situations. An extract from such a task procedure is provided in **CHAPTER 5: ASSESSMENT RECORDS** as the final example of a risk assessment record.

Factors in establishing a safe system of work

17.5 Whether a safe system of work is eventually defined by task safety instructions, a task procedure, a permit to work or a method statement, all relevant factors which might create risks must be taken into account. These will include:

- Equipment and materials used
  - work equipment involved;
  - power sources and the effect of possible power failures;
  - materials used, including their movement or handling;
  - hazardous substances involved directly.
- The work environment
  - access to and egress from the workplace;
  - hazardous substances present in the workplace;

- neighbouring or passing equipment, e.g. vehicles;
- heat, light, dust, fumes and other environmental aspects;
- weather conditions, e.g. wind, ice, rain;

- The work methods
  - frequency to which the task is carried out;
  - variability in the task;
  - complexity of the task;
  - skills required to carry out the task.
- Possible sources of error or problems
  - lack of skill or concentration lapses;
  - fatigue;
  - short cuts likely to be taken;
  - pressure from deadlines or other factors;
  - breakdowns of equipment or emergency situations;
  - external factors, e.g. major distractions.

Development of task procedures

17.6 Detailed task procedures are more likely to be appropriate for higher risk tasks and also for those lower risk tasks which are carried out more frequently. They contain full details of how the task should be performed, providing an important aid to training. They should incorporate all aspects of the task and not just the health and safety issues. These might include:

- The overall purpose of the task;
- Operational instructions
  - in a logical sequence;
  - providing adequate detail.
- Health and safety points
  - highlighting risks, e.g. slippery floors;
  - specifying precautions, e.g. wear safety spectacles.
- Equipment damage prevention
  - detailing risks and precautions, e.g. check oil levels.

- Quality considerations
  - raw materials;
  - product specifications.
- Environment and waste issues.

The following documents may be useful in the preparation of task procedures:

- Engineering manuals or drawings
- Quality control documents
- Training material
- General risk assessment records
- COSHH assessments
- Manual handling assessments
- PPE requirements
  - Other assessments (noise, DSE)
  - Safety rules
  - Accident records
  - Environmental guidelines
  - Waste requirements.

In preparing task procedures (or task safety instructions) it is important to observe and consult with those actually carrying out the task. It may be necessary to go through a couple of drafts before a procedure is produced that is both relevant and contains adequate information on health and safety and other issues. The procedure should be capable of easily being understood and followed by a newcomer to the task. Contentious issues (e.g. the suitability of certain types of PPE) may need to be resolved before the procedure can be finalised. Task procedures should be reviewed periodically to ensure that they are still relevant.

Implementation of safe systems of work

17.7 Employers must ensure that safe systems of work are communicated effectively to those who are expected to follow them and to supervise their use. Formal training is particularly likely to be necessary in relation to the communication of task procedures and also of more formal task safety instructions, e.g. written safety rules. The safe systems of work must also be implemented in practice, requiring appropriate levels of supervision. The effectiveness of this implementation must be monitored – either formally through health and safety inspections and audits or informally by the employer during observations of work activities.

Dynamic risk assessments

17.8 The risk assessment techniques described elsewhere in the book and particularly in **CHAPTER 4: CARRYING OUT RISK ASSESSMENTS** result in what are now commonly described as generic risk assessments. They identify and evaluate the risks normally associated with work activities and the precautions necessary to control those risks to the standard required by legislation.

However, it is impossible for employers to take account in advance of all of the variables involved in work activities and the circumstances and environments in which those activities are carried out. A certain amount of reliance must be placed upon employees to make their own judgements in relation to health and safety. This process is called *dynamic risk assessment* and it is something we can carry out in our daily lives – for example, on every occasion we decide whether or not it is safe to cross the road. It involves:

- identification of risks 'in the field';
- evaluation of the precautions available;
- selection and implementation of the most suitable precautions (or a decision not to go ahead with the task).

Dynamic risk assessment should take place within the framework of a generic risk assessment, i.e. where the risks likely to be associated with the task have been identified and the precautions likely to be necessary have been provided or made available. Those individuals who have received the appropriate level of information, instruction, training and supervision should then be able to carry out dynamic assessments.

Applying this approach to crossing the road, a generic risk assessment should ensure:

- provision of physical controls where appropriate, e.g. light-controlled crossings, zebra crossings, pedestrian central refuges, lighting, barriers, etc.;
- that those who are to cross have the physical capability to do so safely, e.g. eyesight, hearing, walking ability;
- delivery of training in crossing technique, e.g. Green Cross Code, benefits of using crossings;
- supervision of inexperienced people (children in this case) – until they demonstrate the capability to successfully carry out a dynamic risk assessment in practical situations.

Much installation or maintenance work involves work at height using various types of temporary access equipment. Whilst generic risk assessments and project-specific assessments may in many cases have specified what equipment must be used for specific operations, it is impossible to specify everything in advance, and much has to be left to the judgement of individual workers. This is particularly so when staff are working away from their base. In **16.14: DYNAMIC RISK ASSESSMENTS** reference is made to the type of training such staff should receive to carry out such assessments effectively. A generic risk assessment should ensure that:

- suitable temporary access equipment is readily available, e.g. stepladders, extension ladders, tower scaffolds, etc.;
- this equipment is maintained to a satisfactory standard;
- those who are to use it receive training
  - in the safe use of the access equipment;
  - in which types are most appropriate for which circumstances.

Employees trained and equipped in this way should then be able to carry out a dynamic risk assessment of a range of work situations and select the most suitable type of access equipment. In some cases this may involve deciding that none of the equipment available is suitable and requesting alternatives, e.g. full scaffolding or a mobile elevating work platform. It may sometimes be appropriate to decide not to go ahead with the task, e.g. in strong winds or other adverse circumstances.

The 'safe person'

17.9 Some work activities require dynamic risk assessments to be made quite quickly in very dangerous and rapidly changing situations, in order to establish a safe system of work. The emergency services encounter many such situations and fire services, in particular, place reliance on the concept of the 'safe person'. This involves:

- selection of staff with the physical and mental attributes necessary for the work;
- development of procedures for carrying out key tasks;
- provision of equipment (including PPE) necessary to carry out anticipated activities safely;
- training of staff in procedures and the use of equipment;
- the ongoing provision of supervision and information.

The 'safe person' model is one which can be applied elsewhere. Those employees in whom investment has been made in terms of selection, training, procedures, equipment, etc. will be properly equipped to carry out dynamic risk assessments and to establish safe systems of work for themselves.

Permits to work

17.10 In essence a permit to work provides a formal mechanism for ensuring that a dynamic risk assessment of a work situation is carried out by a person competent for the purpose (a 'safe person') in order to ensure that a safe system of work is followed.

Organisations which use permit to work systems will normally require them where:

- The risks involved are high, e.g. work involving high voltage electrical systems, highly flammable liquids or gases.
- Unusual risks are present, e.g. sources of ionising radiation or lasers.
- Complex isolations are necessary to ensure safety, e.g. of power sources, services, material feeds.
- Special precautions are required, e.g. atmospheric testing, use of alarms or special access equipment.
- Personnel are unfamiliar with the environment or risks, e.g. contractors or inexperienced staff.

Further examples of situations where permits to work may be appropriate are:

- work on overhead cranes or crane tracks;
- entry into confined spaces;
- work on potentially dangerous or complex machinery;
- work involving hazardous substances;
- pressure testing;
- work at height;
- excavation work.

Designation of permit situations

17.11 Those situations requiring permits to work should be clearly designated through written task procedures, listings in safety handbooks or rulebooks, the posting of relevant signs and notices or within method statements. It is important that all those likely to be involved (employees and contractors) are fully aware of when permits to work are required. However, it will never be possible to identify all permit situations in advance – staff should always be alert to new tasks which justify the use of a permit to work.

The permit sequence

17.12 For permit to work systems to be effective, isolations and the provision of other precautions prior to work starting (and their subsequent withdrawal) must take place in a strictly controlled sequence. This should be ensured by the manner in which the permit is issued and then cancelled, which would normally follow the sequence below:

- Request for permit issue
  - from the person in charge of the work (or requiring it to be carried out).
- Clearance from person in charge of the area/equipment
  - e.g. production agree that maintenance work can start.
- Isolations made/other precautions taken
  - these must be done or directly supervised by the permit issuer.
- The permit issuer completes the PTW form
  - detailing precautions taken or required, stating any restrictions, e.g. on work activities or areas where work can take place.
- The permit holder signs the PTW form
  - accepting any terms or restrictions.
- The permit is then issued
  - the permit issuer signs the form to issue the permit if satisfied that the permit holder understands and is capable of meeting any requirements in it.

(Normally at least two copies are made – one issued to the permit holder (or put on display), the other retained by the permit issuer or in a central repository).

When work is complete (or must cease)

- *All* permit copies must be brought together.
- The permit holder cancels *all* copies
  - confirming work has terminated, people, equipment, etc. have been withdrawn and the area has been left safe.
- An authorised person cancels *all* copies
  - the permit issuer (or another authorised person) permits the withdrawal of precautions, including isolation removal.
- Isolations are removed and other precautions withdrawn
  - Checks may be made on the general safety of the area/equipment.
- Normal activity may resume.

Permit to work forms

17.13 The detailed design of the form should relate to the types of risks involved and precautions required. The form should include:

- a checklist of risks to be considered;
- some guidance on appropriate precautions;*
- logical places for signatures to issue and cancel the permit;
- clear statements of responsibilities at each stage in this process.

At least two copies of each permit should normally exist – one given to the permit holder or put on display, the other remaining with the permit issuer or at a central repository. Most forms use self-carbon paper, often with colour coding of the different copies. An example of a permit to work form is provided in this section of the chapter.

* In some workplaces standard lists of precautions have been developed for routine tasks controlled by permits to work. Nevertheless the permit issuer must still ensure that the necessary precautions are actually taken and also remain alert for any unexpected risks which may be present and the need for extra precautions.

The permit issuer

17.14 Those issuing and cancelling permits to work should have the following capabilities:

- an appreciation of the purpose of permits to work;
- a clear understanding of the issue and cancellation sequence;
- knowledge and experience of the activities and locations for which they are issuing permits (equipment, risks involved, isolation points, other precautions);
- the ability to carry out, supervise and prove isolations;
- knowledge of other relevant precautions, e.g. gas testing;
- the ability to recognise and deal with problems;
- a positive attitude to health and safety.

Normally there should be a formal system for training, testing and authorising those who are to issue and cancel permits.

Authorisation of permit issuers

17.15 Authorisation should normally be by a well-balanced panel or a very competent individual.

Candidates should be questioned on the theoretical aspects of PTW use:

- the reasons for using PTW systems;
- the issue and cancellation sequence;
- how they would deal with problem situations;

They should then demonstrate their practical knowledge and capabilities:

- the location and use of isolation points;
- the implementation of other precautions.

The panel should issue an authorisation certificate if it is satisfied that the candidate is competent.

The issuing of PTWs must not just 'come with the job'.

PERMIT TO WORK

Serial No. 12345

DETAILS OF PROPOSED WORK			
Intended start date	Intended start time	Expected duration	General clearance is given for this work to take place Signature Position Date

SERVICE ISOLATION CHECKLIST	✓	DETAILS OF ISOLATIONS MADE
ELECTRICITY		
STEAM		
WATER		
GAS		
COMPRESSED AIR		
HYDRAULICS		
OTHER SERVICES		
OTHER POSSIBLE RISKS		**PRECAUTIONS REQUIRED/TAKEN (INC. PPE)**
FLAMMABLE LIQUIDS/GASES		
LACK OF OXYGEN		
CONFINED SPACE ENTRY		
HAZARDOUS SUBSTANCES		
ACCESS/WORK AT HEIGHTS		
MATERIAL FEEDS		
OTHER		
OTHER RESTRICTIONS ON WORK		

ACCEPTANCE	ISSUE	WORK TERMINATION	CANCELLATION
I accept this permit to carry out the work described. I understand the requirements and restrictions described above and all persons under my control will abide by them.	I confirm the isolations have been carried out and other precautions described above have been taken. I am satisfied the permit holder understands the restrictions on the work. THE PERMIT IS ISSUED	The work described above has terminated. All persons and equipment under my control are clear of the area which has been left in a safe condition.	THIS PERMIT IS CANCELLED. All isolations may be removed. Other precautions may be withdrawn. Normal activities may then resume.
SIGNATURE PERMIT HOLDER Date Time	SIGNATURE PERMIT ISSUER Date Time	SIGNATURE PERMIT HOLDER Date Time	SIGNATURE AUTHORISED PERSON Date Time

CDM health and safety plans

17.16 The *Construction (Design and Management) Regulations 1994* (CDM) contain a requirement for a Health and Safety Plan to be prepared for every construction project subject to the Regulations. (The definition of 'construction work' used in the Regulations is so broad that many types of maintenance and installation work also fall within the scope of the *CDM Regulations 1994* – see **REF. 1**.)

The purpose of the plan is to ensure that risks expected to be involved in the project are identified in advance, together with the precautions necessary to control those risks. The framework of the plan must be prepared initially by the 'Planning Supervisor' (a person or organisation appointed by the 'Client').

The 'Principal Contractor' for the project must not be appointed until the 'Client' (on the advice of the 'Planning Supervisor') is satisfied that they are competent to implement the precautions identified in the plan and that they will allocate adequate resources for health and safety. Responsibility for the detailed development and implementation of the plan then passes to the 'Principal Contractor' who also has responsibilities in respect of the competence and health and safety standards of other contractors and self-employed persons working on the project.

Detailed HSE guidance is available on the Regulations themselves, the definitions and responsibilities of the various duty holders and on the content of Health and Safety Plans (see **REFS 1–3**).

The *CDM Regulations* 1994 require the use of risk assessment techniques in the preparation of the Health and Safety Plan and then the application of safety management techniques in its implementation.

The 'Principal Contractor' and other contractors involved in the project should already have *generic risk assessments* relating to their work – the risks normally associated with the work and the precautions available or normally applied. The Health and Safety Plan involves a form of *dynamic risk assessment* in relation to this project including consideration of:

- health and safety standards to be met in the work;
- co-ordination of the contractors involved;
- overall site PPE requirements;
- provision of services, facilities and equipment;
- special requirements of the site or its location;
- special or unusual requirements of the project.

The accompanying Health and Safety Plan checklist was developed primarily to assist those acting as 'Planning Supervisors' or 'Principal Contractors' for engineering and minor building projects in established workplaces such as factories or offices. The content of Health and Safety Plans for other types of work, e.g. major new build or demolition projects, will undoubtedly be different. However, the checklist provides at least an illustration of the types of issues which must be addressed in preparation of the plan.

Construction (Design and Management) Regulations 1994

HEALTH AND SAFETY PLAN CHECKLIST

This checklist is intended to assist those fulfilling roles as Planning Supervisors or Principal Contractors under the *CDM Regulations 1994* but may also help clients in identifying important issues relating to health and safety.

Planning supervisors should use it to identify risks and other health and safety issues of relevance to the project and to outline what precautions and other arrangements are likely to be necessary.

Principal contractors will need to provide more detail of what precautions actually have been or will be taken. Where appropriate, cross references should be made to other documents (**see SECTION 2 OF TABLE: RELEVANT DOCUMENTS**) and additional detail may be provided on separate sheets.

For small projects the careful completion of this form should provide adequate information but for larger projects, especially those involving significant risk, much more information is likely to be required – including reference to relevant risk assessments, method statements, etc.

1	PROJECT DETAILS	
1.1	Project Title	
1.2	Expected Commencement Date	
1.3	Predicted Duration	
1.4	Project Description	
2	RELEVANT DOCUMENTS	Give details of relevant documents and standards
e.g.	Project specifications Drawings HSE Guidance BS or CE specifications IEE standards Existing health and safety files	
3	CONTRACTORS	Contractors must be competent and adequately resourced
3.1	Principal Contractor	
3.2	Other Contractors	
4	COMMUNICATIONS	What arrangements will be made for
4.1	Monitoring design work	
4.2	Considering design change implications	
4.3	Project review meetings	
4.4	Inducting contractors' employees	
4.5	Checking contractors' employee training	
4.6	Monitoring the work location	
5	PERSONAL PROTECTIVE EQUIPMENT	What will the site requirements be for
5.1	Safety footwear	
5.2	Safety helmets	
5.3	Clothing (inc. high visibility)	
5.4	Eye protection	
5.5	Hearing protection	
5.6	Gloves	
5.7	Other PPE	
5.8	Signs	
6	SITE SERVICES AND EQUIPMENT	Arrangements for their provision and maintenance
6.1	Electrical power	
6.2	Water supplies	
6.3	Compressed air	
6.4	Lighting	
6.5	Plant or equipment	
6.6	Other	
7	TRAFFIC AND TRANSPORT	
7.1	Traffic routes	
7.2	Parking	
7.3	Speed limits	
7.4	Headroom	
7.5	Emergency access routes	
7.6	Separation from moving traffic	

8	PEDESTRIAN SAFETY	
8.1	Access to and within site	
8.2	Access for work at heights	
8.3	Protection of openings	
8.4	Protection from falling materials	
9	SECURITY/SEGREGATION	
9.1	Segregation from other activities	
9.2	Perimeter security	
9.3	Warning signs	
9.4	Security attendance/patrols	
10	HAZARDOUS SUBSTANCES	
10.1	Presence within site	
10.2	Use in the project	
10.3	Storage or disposal issues	
11	OTHER SPECIAL ISSUES	
11.1	Buried or overhead services	
11.2	Stability of other structures	
11.3	Need for Permits to Work	
11.4	Commissioning	
12	FIRE	
12.1	High risk materials/activities	
12.2	Exit routes	
12.3	Evacuation procedures	
12.4	Fire fighting equipment	
13	FACILITIES AND ARRANGEMENTS	
13.1	Material storage	
13.2	Waste storage and disposal	
13.3	Welfare facilities	
13.4	First Aid	

Signed Date

Date passed to Principal Contractor (if relevant)

Method statements

17.17 A safety method statement is a description of how risks will be controlled or managed in relation to a specific task or activity. It may be incorporated within an overall method statement which describes how the whole task will be performed, i.e. including detailed work methods and specifications of equipment and materials to be used. Method statements are increasingly being required for construction-related projects, with the safety method statement providing an important means of complying with the *CDM Regulations 1994*.

The health and safety plan should normally cover risks in relation to the project as a whole. The safety method statement will normally deal with a task, an activity or a specific health and safety issue identified in the plan and go into much more detail. It will be based on a risk assessment of the relevant task or activity. The content of a method statement might include:

- Identification of key personnel
  - for overall control and specific operations.
- Training requirements
  - e.g. for use of cranes, fork-lifts, etc., or the testing or commissioning of equipment.
- Access requirements
  - for vehicles and pedestrians;
  - access equipment needed;
  - emergency access issues.
- Equipment requirements
  - size, weight, type;
  - power rating, certification;
  - location, stability.
- Site requirements
  - traffic considerations;
  - security/protection of the public or other workers.
- Materials involved
  - storage, transportation, handling, disposal;
  - hazardous substances.
- Work sequencing
  - need for temporary precautions;
  - key scheduling issues.
- Environmental considerations
  - e.g. wind speed limits;
  - rain, temperature.
- PPE requirements
- Other precautions
  - barriers, signs, etc.;
  - fire fighting equipment;
  - detection equipment;
  - rescue equipment.

References

(All HSE publications)

17.18

1	HSG 224	Managing health and safety in construction: *Construction (Design and Management) Regulations 1994*. Approved Code of Practice and Guidance (2001)
2	CIS 42	*Construction (Design and Management) Regulations 1994*. The pre-tender stage health and safety plan (1995) – free leaflet
3	CIS 43	*Construction (Design and Management) Regulations 1994*. The health and safety plan during the construction phase (1995) – free leaflet

18 Assessing and managing risk – can you afford not to?

In this chapter:

Introduction	**18.1**
The Costs of Accidents at Work	**18.2**
Some high-profile examples	**18.3**
Piper Alpha	**18.3**
Lanarkshire gas explosion	**18.4**
Heathrow tunnel collapse	**18.5**
Hatfield rail crash	**18.6**
The benefits of assessing and managing risk well	**18.7**
A safe, healthy and well-motivated workforce	**18.7**
A favourable reputation	**18.8**
A smooth operation	**18.9**
The potential costs of failure	**18.10**
Fines and legal costs	**18.10**
Insurance costs	**18.11**
Supply-chain pressure	**18.12**
Loss of customer confidence	**18.13**
Staff absence	**18.14**
Repair, replacement and rectification	**18.15**
Business interruption	**18.16**
Administrative and other costs	**18.17**
References	**18.18**

Introduction

18.1 The effective management of health and safety should be an integral part of the management of any business, whatever its size or work activities. Risk assessment and the implementation of appropriate control measures (through the application of the 'management cycle' (see **8.3: THE MANAGEMENT CYCLE**)) must inevitably form an essential part of any health and safety management programme. This chapter seeks to demonstrate the benefits to businesses of assessing and managing risks effectively and the costs of failing to do so.

The Costs of Accidents at Work

18.2 In 1993 the HSE published a booklet entitled *The Costs of Accidents at Work* (**REF. 1**). This was based on case studies carried out by the HSE's Accident Prevention Advisory Unit (APAU) on organisations from five different sectors of industry. Their research aimed to develop on earlier work by Heinrich, Bird and others (see **8.18: ACCIDENT RATIO STUDIES**), which had shown that for every major or lost time injury there were many more minor injuries and a far greater number of non-injury incidents.

The APAU demonstrated through their case studies that the true costs of injuries and incidents were far greater than most people realised and that most of those costs were ones that the businesses were not insured for. The ratios between insured and uninsured costs for four of the five organisations are shown below.

	Insured: Uninsured costs ratio
Construction site	1:11
Creamery	1:36
Transport company	1:8
North Sea oil production platform	1:11

(Full cost details were not collected for the fifth study, in an NHS hospital.)

Accidents were estimated to cost:

- the construction contractor 8.5% of its tender price (on an £8 million tender);
- the transport company 37% of its annualised profits; and
- the NHS hospital 5% of its operating costs.

Some of the costs of accidents can be seen directly (e.g. compensation claims, costs of repairs), whereas others have a more indirect effect on the business (e.g. loss of customer goodwill). Even many direct costs (e.g. the costs of minor repairs) are not routinely quantified by most organisations. The accompanying figure (below) demonstrates the make-up of insured/uninsured and direct/indirect costs of accidents, and several of these costs, together with the benefits of good risk management, are explored in greater detail later in the chapter.

The costs of accidents

INSURED

	DIRECT	INDIRECT
INSURED	• Employers' liability claims • Public liability claims • Major damage • External legal costs	• Major business interruption • Product liability
UNINSURED	• Fines • Sick Pay Minor damage Lost or damaged product Clear-up costs	Minor business interruption Prohibition notice interruption Customer dissatisfaction (product delay) First aid and medical attention Investigation time and internal administration Diversion of attention/spectating time Loss of key staff Hiring/training replacement staff Hiring replacement equipment Lowered employee morale Damage to external image

UNINSURED

- Only these costs are capable of being quantified without considerable effort

(This is an expanded version of a diagram in the HSE booklet *The Costs of Accidents at Work* (**REF. 1**).)

Some high-profile examples

Piper Alpha

18.3 The explosion and fire in 1988 on the North Sea oil platform resulted in the loss of 167 lives and is estimated by the HSE (**REF. 1**) to have cost over £2 billion, including £746 million in direct insurance payouts.

Lanarkshire gas explosion

18.4 A new record fine for health and safety offences of £15 million was set in 2005 when Transco were convicted of a breach of *Section 3(1)* of *HSWA*. The prosecution followed a large gas explosion in 1999 which killed a couple and their two children living in a bungalow in Larkhall, Lanarkshire. Transco were criticised for lack of effective inspection and maintenance arrangements and inaccurate records of pipes – the ductile iron gas main was badly corroded.

Heathrow tunnel collapse

18.5 No one was injured when a tunnel which was to form part of the Heathrow Express rail link to central London collapsed during its construction in 1994, but the incident was described in court as one of the biggest near misses in years. Fines and costs totalling £1.4 million were imposed on the two contractors held responsible. The associated costs of the disruption to flights at a major international airport would be massive.

Hatfield rail crash

18.6 In 2005 Balfour Beatty were fined £10 million and Network Rail (formerly Railtrack) £3.5 million for their parts in the rail crash in 2000 at Hatfield, in which four people died when a train was derailed as a result of poor standards of inspection and maintenance of the track. Apart from the fines themselves, the incident had already had a profound effect on the rail industry. Major disruption was caused to the rail network resulting in huge payments having to be made to compensate the train operators and many of the travelling public. The event played a significant part in the eventual demise of Railtrack, and the relationships between their successor body and the contractors actually carrying out maintenance and repair work on the track are to be totally restructured.

The benefits of assessing and managing risk well

A safe, healthy and well-motivated workforce

18.7 Apart from the humanitarian aspects associated with not causing injury or illness to employees (and others), there are many other benefits from maintaining high standards. HSE's *Successful Health and Safety Management* (**REF. 2**) refers to the importance of the health and safety culture, which it defines as:

> the product of individual and group values, attitudes, perceptions, competencies and patterns of behaviour that determine the commitment to, and the style and proficiency of, an organisation's health and safety management. Organisations with a positive safety culture are characterised by communications founded on mutual trust, by shared perceptions of the importance of safety and by confidence in the efficiency of preventive measures.

Put more briefly, the safe way becomes 'the way we work here'.

There are many more indirect benefits from establishing a positive safety culture. Employees will be more motivated to care about the interests of an employer whom they perceive as caring for them, particularly when it comes to 'going the extra mile' in the course of their work. They will also assist in developing the reputation of their employer as a 'good firm' which can have considerable effects within the wider community where there are potential customers, business partners and future employees (see **18.8: A FAVOURABLE REPUTATION** below). Many experienced managers and health and safety professionals can provide anecdotal evidence of situations where an improving health and safety record went hand in hand with a developing pride in the company and increased productivity figures.

The commitment of the workforce is best achieved through involving them in health and safety matters. The importance of doing this during the risk assessment process was stressed in earlier chapters (see **4.11: DISCUSSIONS** in particular). Consultation with employees should take place on an ongoing basis – both formally (see **8.21: HEALTH AND SAFETY COMMITTEES**) and informally with individual employees and their representatives. Precautions are much more likely to be improved through involvement rather than imposition.

A favourable reputation

18.8 We are all influenced by reputation, whether in the brands we buy or in the shops and services that we use. There are also many instances where a favourable reputation, built up over many years, can be destroyed in a very short time, sometimes by a single incident.

Good reputations in health and safety-related matters are established through the words and actions of employees, the experiences of customers or clients of the business and their interactions through word of mouth 'networking'. This process will undoubtedly be assisted by the ready availability of suitable documents detailing the organisation's health and safety policy, recording its risk assessments and providing relevant method statements.

However, all this can be undermined by a major accident, a prosecution or even the behaviour of an individual employee. Risk assessment is important but so also is risk management – a failure to manage risk effectively at the point where work is being carried out can have a catastrophic effect on a carefully established reputation.

The reputation of a business with enforcing authorities is also an important consideration. The HSE programme of routine visits to workplaces has for many years been influenced by their perception of how well the employer is managing health and safety – the poorer performers being visited more frequently. In addition a failure in health and safety management is more likely to be regarded sympathetically by HSE or local authority Inspectors (and thus less likely to result in a prosecution or other enforcement action) if they consider it an isolated breakdown in an otherwise well-managed organisation.

A smooth operation

18.9 A business that has carried out thorough risk assessments will be much more aware of the potential for things to go wrong and result not only in accidents and ill health but also in damage and disruption. The application of good safety management principles through the management cycle (see **8.3: THE MANAGEMENT CYCLE**) should both reduce the numbers of accidents and incidents occurring and reduce the unfavourable consequences of those that do occur.

Regulation 8 of the *Management Regulations 1999* requires the establishment and implementation of 'appropriate procedures to be followed in the event of serious and imminent danger to persons at work' (see **2.13: PROCEDURES FOR SERIOUS AND IMMINENT DANGER AND FOR DANGER AREAS (REGULATION 8)**). Such emergency procedures should be closely integrated with contingency planning to minimise *all* of the adverse effects which might result from foreseeable incidents.

Well-managed businesses minimise the opportunities for surprises to injure their employees or damage their interests.

The potential costs of failure

Fines and legal costs

18.10 Prosecutions for more serious health and safety offences are heard in the Crown Court, which may impose unlimited fines and, for some specified offences, custodial sentences. The highest fine ever imposed for a single offence increased in 2005 to £15 million, in the case resulting from a gas explosion in Lanarkshire. In the same year a fine of £10 million was imposed on an engineering company as a result of the Hatfield rail crash. With further serious incidents still due to come to court this figure may well increase soon.

Most prosecutions are dealt with by magistrates courts who can currently impose a maximum fine of £20 000 for offences under *HSWA 1974* – an amount which could have a major impact on a small business. It is, of course, open to the enforcing authorities to bring multiple charges against employers should the circumstances justify this.

Insurance costs

18.11 The holding of Employers' Liability Insurance (in respect of injury to employees) is a compulsory legal requirement for most employers. (Some public bodies are exempt and there are also exceptions in relation to the employment of close family members.) Motor vehicle insurance is also a statutory legal requirement, constituting a major expense for many businesses. Whilst insurance for liabilities to members of the public and third parties (including customers or clients) is not compulsory, it is nevertheless an essential investment for any prudent employer or self-employed person. Indeed, contractors without such insurance will find it almost impossible to get work from other businesses – see **18.12: SUPPLY-CHAIN PRESSURE** below. Other commonly held types of insurance relate to product liability and damage to buildings, equipment, stock [illegible] raw materials.

Most forms of insurance follow a similar pattern to the motor vehicle insurance with which most readers will be familiar. Those employers with a poor claims record will find it increasingly difficult to get insurance and only then at higher premiums. Many insurers will increase premiums because of claims which are known to be in the pipeline, even though no payment has yet been made and liability may not even have been admitted.

Another parallel with motor vehicle insurance is that policyholders are required (or encouraged by premium levels) to meet the first part of any claim themselves, i.e. the insurance is only activated by claims above a certain value. Consequently much of the cost of 'routine' minor damage, e.g. to premises or equipment, is borne by the employer rather than the insurer.

Insurers and insurance brokers are pro-actively seeking evidence that both prospective and existing customers have carried out effective risk assessments. Those employers who cannot produce risk assessments will find it increasingly difficult (or more expensive) to obtain necessary insurance.

As a result of recent changes to legislation, civil actions can now be brought by workers against their employers for breaches of the *Management of Health and Safety at Work Regulations*. It would be difficult for an employer to claim that a particular risk was not foreseeable when no risk assessment of the activity had been carried out.

Those who carry out risk assessments but fail to provide effective management systems to ensure that the necessary precautions are in place are possibly even more vulnerable at civil law. Identifying that precautions are necessary and then failing to implement them is providing an open goal for claimants and their legal representatives to shoot into.

Supply-chain pressure

18.12 Since the introduction of *HSWA 1974*, many court decisions, most notably those involving Swan Hunter Shipbuilders and Associated Octel, have emphasised the responsibilities of employers for the activities of employees of other organisations. The structure of *HSWA 1974* and much subsidiary legislation is such that responsibilities overlap between employers rather than being neatly apportioned between them. Employers must do more than simply not turn a blind eye to the obvious health and safety failings of those with whom they come into contact: they must often take a pro-active interest in the health and safety standards of others.

In recent years a growing number of larger companies, local authorities and other public bodies have put into practice increasingly formalised procedures for checking the health and safety standards of contractors wishing to work for them. This process has been accelerated by the demands of the *Construction (Design and Management) Regulations 1994 (CDM Regulations 1994)* which require clients to satisfy themselves (via their planning supervisors) that potential principal contractors are capable of dealing with the health and safety issues associated with projects. The Regulations also place responsibilities on principal contractors in respect of their sub-contractors. Consequently contractors are frequently required to provide details of their health and safety policies and generic risk assessments together with risk assessments and/or method statements for specific projects or activities. Many clients also take an extremely hands-on approach in policing the work of contractors on their premises.

The importance of supply-chain pressure was stressed in the Strategy Statement *Revitalising Health and Safety* published in June 2000 by the Health and Safety Commission and the Department of the Environment, Transport and the Regions (DETR) – then the parent government department of the HSC/HSE. This states:

> All public bodies must demonstrate best practice in health and safety management. Public procurement must lead the way on achieving effective action on health and safety considerations and promoting best practice right through the supply chain.

Consequently in order to gain new business or to retain existing work, employers will increasingly have to demonstrate that they have carried out risk assessments and also have effective management systems in place to ensure that necessary precautions are implemented.

Loss of customer confidence

18.13 The possibility of customer confidence being undermined in relation to health and safety standards was examined earlier in the chapter (see **18.8: A FAVOURABLE REPUTATION**). Customer confidence may also be lost through a failure to deliver goods or services on schedule or to the required standard. This failure may be attributable to an accident or incident – damage to goods, equipment or premises, or an injury to a key member of staff.

No matter how sympathetic the customer may be to the supplier's position, they have their own interests to consider. Customers are likely to seek alternative sources of supply, particularly if problems are likely to continue or recur. Many businesses affected by serious fires never re-open. Even if they could afford to, they would find that they had lost their market.

Staff absence

18.14 The potential effects of the absence of key staff were considered in the previous section (see **18.13: LOSS OF CUSTOMER CONFIDENCE**). Even where staff can be replaced effectively this is likely to involve extra costs – overtime payments and premium rates to outside agencies as well as administrative time taken up with arranging the necessary cover. This would be in addition to sick pay made to staff who are absent due to injury or occupational ill health.

Repair, replacement and rectification

18.15 As stated previously, many of the costs associated with the repair or replacement of damaged equipment, premises or materials or the rectification of damaged products will be below the levels at which insurance cover comes into operation (see **18.11: INSURANCE COSTS**). For example a minor collision of a vehicle with a building could result in damage to the vehicle itself, the building, equipment housed within the building as well as materials or product being carried by the vehicle.

Business interruption

18.16 The potential effects of business interruption on customer confidence were considered earlier (see **18.13: LOSS OF CUSTOMER CONFIDENCE**), as were the costs of staff absence. However, both business interruption and lost staff time may also occur as a secondary result of poor safety management.

This might be due to the enforcing authorities using their powers to issue a 'prohibition notice' – whether following an accident or simply because of concerns about serious potential dangers. The author recalls one situation where approximately 500 employees had to be diverted from production work for six hours because of a prohibition notice and he is also aware of other cases where key pieces of equipment were taken out of action for several days.

Major interruptions may also be caused by industrial action being taken by workers, either due to general dissatisfaction about health and safety standards or as a reaction to a specific accident or incident. In some cases such industrial action may also have serious adverse effects on other employers, e.g. where a principal contractor struggling to meet a project deadline faces a walkout from staff employed by a sub-contractor. This provides another strong justification for the increasing supply-chain pressure on contractors referred to previously (see **18.12: SUPPLY-CHAIN PRESSURE**).

Administrative and other costs

18.17 Accidents and damage incidents can disrupt the smooth running of businesses in many other ways. Damaged equipment has to be recovered and debris needs to be cleared up. There will be costs associated with the provision of first aid treatment, in accompanying injured staff for hospital treatment or taking them home, and in passing on information to their next of kin. Where serious accidents or incidents have occurred, much time will be wasted by others simply coming to have a look – as exemplified by the 'rubbernecking' effects on traffic on the opposite carriageway in the case of road accidents.

Additional administrative time will be involved in investigating the causes of the accident or incident (hopefully this will be put to good effect in preventing recurrences), dealing with enforcing authority inspectors during their investigations and in providing information for insurance companies, solicitors and others involved in processing claims for damages.

References

18.18

1	HSG 96	The costs of accidents at work (HSE, revised in 1997)
2	HSG 65	Successful health and safety management (HSE, revised in 2000)
3		Revitalising Health and Safety (DETR/HSC 2000)

19 Looking ahead

In this chapter:

Risk assessment is here to stay **19.1**

Possible changes **19.2**

Integration of requirements **19.2**
CDM/CHSW Regulations **19.3**
Control of Asbestos Regulations **19.4**

Widening the concept of work-related risks **19.5**

Stress at work **19.6**
Work-related violence **19.7**
Work-related travel **19.8**

References **19.9**

Risk assessment is here to stay

19.1 Reference was made in CHAPTER 1: INTRODUCTION to some of the recommendations of the Robens Committee, made in 1972, which eventually resulted in the *Health and Safety at Work etc Act in 1974 (HSWA 1974)*. One of the themes of the Robens report was that of 'self-regulation' – that employers should address all of the risks involved in their activities rather than just those for which there were specific legal provisions.

To that end, *HSWA 1974* contained a number of general obligations of employers and the self-employed both towards their own employees and to others not in their employment, who might be affected by their work activities. These requirements were qualified by the phrase 'so far as is reasonably practicable'. As was explained in CHAPTER 1: INTRODUCTION, this in effect required employers to carry out a risk assessment in order to identify foreseeable risks and determine what were or were not reasonably practicable precautions. CHAPTER 17: RISK ASSESSMENT RELATED CONCEPTS also explained how the process of risk assessment is fundamental in determining 'a safe system of work' – another basic obligation under *HSWA 1974*.

This progress towards 'self-regulation' was further formalised by the requirements contained in the *Management Regulations 1999* and other specific codes of Regulations for risk assessments to be made and (in most cases) to be recorded. In effect the *Management Regulations 1999* oblige employers to carry out a risk assessment every time a new statutory requirement applying to their activities is introduced. Risk assessments must also be reviewed whenever any significant change is made to any aspect of those work activities.

Revitalising Health and Safety, the Strategy Statement published in June 2000 by the Department of the Environment, Transport and the Regions (DETR), the then parent Department of the Health and Safety Executive (**REF. 1**), also referred to 'self regulation'. One of the 10 key points in the Strategy Statement was the need to cultivate 'a more deeply engrained culture of self regulation', particularly in small businesses. Risk assessment by employers must be a fundamental part of that self-regulation.

Possible changes

Integration of requirements

19.2 One change that has been discussed for some time is the integration of all risk-assessment requirements under a single statutory provision rather than the present situation where requirements are contained in several different codes of Regulations. Whilst this might have the benefit of simplifying the law, it would have little tangible impact in practice. As this book should have demonstrated, the techniques involved in carrying out one type of risk assessment are very similar to those necessary for the other types.

CDM/CHSW Regulations

19.3 Consultation took place during 2005 on revision of the *Construction (Design and Management) Regulations 1994*. It was intended that the revised version would incorporate the parts of the *Construction (Health, Safety and Welfare) Regulations 1996* which were not replaced by the *Work at Height Regulations 2005*. Whilst not expected to contain any new risk assessment requirements, the new regulations are likely to make changes affecting Health and Safety Plans (see **17.16: CDM HEALTH AND SAFETY PLANS**) and other changes which may affect risk assessments for construction work.

Control of Asbestos Regulations

19.4 Consultation opened during 2005 on proposals for new 'Control of Asbestos Regulations' and an associated Approved Code of Practice. These would replace the *Control of Asbestos at Work Regulations 2002*, the *Asbestos (Licensing) Regulations 1983* and the *Asbestos (Prohibition) Regulations 1992* with a single set of regulations. The requirements of the new regulations may affect the assessment of risks from asbestos in premises (see **CHAPTER 15**).

Widening the concept of work-related risks

19.5 The boundaries of what constitutes a work-related risk are constantly being pushed back, particularly in the field of civil litigation. The last decade has seen increasing interest and legal activity in relation to newer areas of risk such as stress, work-related violence and travel for work purposes. The extension of civil liability referred to in **19.6** below is likely to accelerate this process.

Stress at work

19.6 Following the successful landmark case of *Walker v. Northumberland County Council* in 1995 (**REF. 3**), there have been an increasing number of civil claims against employers in respect of work-related stress, many of which have been successful. The TUC statistics recorded a total of 6428 new stress cases brought during 2001. In February 2002 the Court of Appeal delivered judgments on appeals by four employers against county court judgments on stress claims, upholding three of the appeals (**REF. 3**). However, the Court also laid down guidelines for future claims arising from work-related stress. These consisted of 16 'practical propositions', some of which could only be determined through a process of risk assessment .

For example, reasonable foreseeability of harm includes consideration of aspects such as the nature and extent of the work, whether the work is particularly intellectually or emotionally demanding for the individual employee, and whether others doing the same job are suffering harmful levels of stress. There would be a breach of duty only if the employer has failed to take steps that are reasonable in the circumstances, bearing in mind the magnitude of the risk of harm occurring, the gravity of that harm, the cost and practicability of preventing it and the justification for running the risk – all factors which should be dealt with in the risk-assessment process.

Apart from the possibility of civil claims in respect of work-related stress there are other costs associated with the problem. Research commissioned by the HSE has indicated that

1 around half a million people in the UK experience work-related stress at a level they believe is making them ill;

2 upto 5 million people in the UK feel 'very' or 'extremely' stressed by their work;

3 12.8 million working days were lost to stress, depression and anxiety in 2004/5.

The HSE define stress as 'the adverse reaction people have to excessive pressure or other types of demands placed on them'. Stress is people's natural reaction to excessive pressure and in some circumstances can be a good thing, improving performance. However, where excessive pressure continues for some time it can lead to both mental and physical ill health.

There are many causes of stress outside work, including death or illness of a close relative or friend, moving house or even marriage and holidays, and these types of stress will obviously also have an impact on employees whilst they are at work. However, stress may also be caused or made worse by working conditions, work activities or work relationships.

Working conditions causing stress can include:

- excessive or distracting noise;
- poor lighting;
- too high or too low a temperature; and
- inadequate desk space or general congestion.

Work activities and the way they are organised may be stressful due to factors such as:

- unrealistic work targets;
- tasks beyond the capabilities of the individual;
- under-utilisation of a person's skills;
- boring or repetitive tasks;
- constant changes in activities or organisation;
- lack of direction or decision-making;
- social or professional isolation;
- traumatic or violent incidents;
- excessive or constantly changing hours of work; and
- job insecurity.

Relationships at work can be stressful because of:

- clashes or tensions with supervisors or colleagues;
- bullying, harassment or discrimination; and
- threats or abuse from customers or clients.

Physical conditions which may cause stress should be identified through observation (see **4.10: OBSERVATION**) and then rectified by the effective application of the principles of risk assessment described in **CHAPTER 4: CARRYING OUT RISK ASSESSMENTS**. However, the organisation of work activities and interpersonal relationships are much more closely related to the culture prevailing within the organisation.

Stress will be much less of a problem in organisations which have:

- clear policies and procedures which are implemented effectively (especially in key areas such as bullying, harassment, discrimination, grievances, etc.);
- effective communications and consultation (in relation to business objectives and, particularly, proposed changes);
- opportunities for employees to contribute (on how work is planned and organised);
- clearly defined rules and appropriate training (including training in interpersonal skills and stress awareness);
- support for staff affected by stress (access to counselling services, stress management training, etc.).

The HSE have developed 'Management standards for work related stress' (available on their website) to help employers, employees and employee representatives to work together to manage stress sensibly and minimise the impact of work-related stress on business and, of course, on workers. Several HSE publications also provide guidance on managing stress (**REFS 4–10**). Apart from their legal obligations to identify stress risks and introduce appropriate control measures, there are many cost benefits to employers from managing stress effectively – through reduced sickness absence, lower staff turnover and improved staff performance.

Work-related violence

19.7 Regrettably workers in many occupational areas are increasingly subject to actual or threatened violence. Indeed the possibility of violence is often a contributing factor in work-related stress cases.

The HSE define work-related violence as:

> any incident in which a person is abused, threatened or assaulted in circumstances relating to their work.

This is a particularly widely drawn definition and means that threats taking place outside working hours and away from the workplace still constitute work-related violence. The 2000 British Crime Survey referred to nearly 1.3 million violent incidents involving people at work in England and Wales during 1999. These consisted of 634 000 physical assaults and 654 000 cases of threats or threatening behaviour. The latter figure is probably considerably below the true level and does not take into account incidents involving abuse, which are included in the HSE definition.

Acts of 'non-consensual physical violence' resulting in death, major injury or incapacity for normal work for three or more days are now reportable under the *Reporting of Injuries, Diseases and Dangerous Occurrences Regulations 1995 (RIDDOR)*.

Risks are generally highest for those in front-line contact with members of the public, particularly the health care, education, social services, retail, public transport and hospitality sectors. However, many other workers are also at risk, although their employers may well be unaware of this, particularly if there have been no actual instances of assault.

Employers first need to establish whether there are problems with work-related violence. This can be done through informal consultation with staff, by asking staff to fill in relevant questionnaires or through discussions with employee representatives. Where problems are suspected or are known to exist, employees, supervisors and managers should be encouraged to report all relevant incidents – including abuse and threats.

Some types of violence might occur on the employer's premises. These might involve:

- handling cash or valuables – e.g. banks, building societies, post offices, security staff, cashiers, retail and bar staff;
- dealing with the public – e.g. complaints and enquiry desks, retail and hospitality workers, education;
- dealing with high-risk people – e.g. prison and court staff, residential care work, hospital casualty units.

However, employers must also take into account the risks of violence to their staff whilst working away from their own premises, including work in the community. Some potentially high-risk activities include:

- work by the emergency services – fire, police and ambulance staff;
- other enforcement, inspection and control work – e.g. environmental health officers, traffic wardens, ticket inspectors, planning officers, park rangers, security staff;
- domiciliary work – e.g. home carers, social services, charity workers, service engineers;
- investigative work – e.g. journalists, photographers, private detectives;
- door-to-door sales and delivery work – (such staff also often carry cash).

The risks may not just be due to the work activities themselves but may also relate to the area where, or the time at which, the activity is carried out. Many inner city areas and rough estates can present risks at any time of day, as can derelict buildings or land. These risks will often increase during the hours of darkness and particularly at times when pubs and clubs are closing.

Some staff are often required to work and to travel home at anti-social times, e.g. hospitality and entertainment workers, cleaning staff and those required to lock up premises staying open late. Designated key-holders are also vulnerable when they are called out to deal with an activated intruder alarm or an actual break-in.

As with other types of risk assessment, once the nature and extent of the risk has been identified, an assessment must be made of the adequacy of the existing control measures. These generally fall into two categories:

- the design and equipping of the workplace itself; and
- the establishment of safe systems of work.

Workplaces may be made more secure and safe by:

- restricting public access – e.g. by locked doors (possibly with keypad or swipe card control), separate staff entrances, gates or barriers;
- separating staff from the public – e.g. by screens or barriers, wide counters, raised floors on the staff side, separate staff entrances;
- reducing stress for the public – e.g. through clear signs, pleasant designs for reception areas, avoiding queuing situations;
- keeping cash secure – e.g. use of safes (preferably with time delays), drop safes on tills and, where relevant, signs indicating that staff do not have access to cash;
- clear visibility outside buildings, including parking areas – e.g. good lighting and lines of sight (bushes, etc. kept pruned), closed-circuit TV (clearly evident);
- emergency alarm devices – e.g. fixed alarms at high-risk locations, portable alarms where appropriate.

Risks can be reduced or even avoided altogether by establishing safe systems of work, supported by the provision of relevant equipment. Safe systems may be necessary for some risks on the employer's own premises but are particularly important for those who are potentially at risk elsewhere. They are likely to involve some of the measures outlined below:

- reducing cash holdings – e.g. by encouraging use of direct debits, cheques, credit cards, pre-payment systems, etc. and banking cash frequently;
- planning how cash will be transported – e.g. providing suitable transport, varying routes;
- checking client credentials – e.g. by obtaining relevant information from referral agencies, maintaining accessible records of high-risk clients;
- choosing meeting locations – avoiding encountering high-risk or unknown people in remote places or their own homes – e.g. by meeting in public buildings or other suitable places such as cafés;
- providing communication equipment – e.g. mobile phones, radios, attack alarms;
- keeping track of vulnerable staff – e.g. through use of mobile phones (and taking follow-up action if they don't report in or return when expected);

- providing secure transport – e.g. taxis, lifts from colleagues – for travel in high-risk areas or at high-risk times;
- accompaniment by colleagues or others – where the level of risk justifies this;
- selecting suitable staff for the task – some staff may be more vulnerable in certain situations, because of their gender, age, ethnic origin and particularly because of their lack of experience.

The *Management Regulations 1999* require employers to establish appropriate emergency procedures (see **2.13: PROCEDURES FOR SERIOUS AND IMMINENT DANGER AND FOR DANGER AREAS (REGULATION 8)**. Threatened or actual violence is certainly 'serious and imminent danger' as far as the vulnerable employee is concerned. Where the risks merit it, employers must establish suitable procedures for:

- response to alarms, requests for assistance, coded messages, etc.;
- investigation when staff don't report in or return;
- the involvement of security staff or the police when necessary.

In particularly high-risk situations consideration might need to be given to the possible need for mediation or rescue.

Whilst employers must carry out generic risk assessments of the possible exposure of their staff to violence, there will also be a need for some staff to themselves make dynamic assessments of the risks associated with their work activities on a day-to-day basis (see **17.8: DYNAMIC RISK ASSESSMENTS**). Such staff must receive appropriate training so that they are able to identify the risks which may be involved with a particular activity and also so that they are aware of the equipment available to them (e.g. mobile phone, attack alarm), systems of work which might be adopted (e.g. accompaniment) and any relevant emergency procedures (see **17.9: THE 'SAFE PERSON'**). Training in techniques for dealing with potentially violent situations is also likely to be necessary. Much pioneering work has been done in this area by the Suzy Lamplugh Trust who have a range of training material available (**REF. 11**).

Where incidents of violence do occur, these should be investigated and the risk assessment should be reviewed and, if necessary, revised. In serious cases there may also be the need for counselling of the victims and for time-off for victims to recover, and it may even be appropriate for employers to provide legal assistance to institute civil actions on behalf of the victims against the perpetrators.

However, many organisations are still not properly addressing the possibilities of their staff being subject to violence. Their failure to assess the risks and to introduce reasonable precautions not only increases the chances of a violent incident occurring, but also leaves them more vulnerable to criminal prosecutions and civil actions in respect of such incidents.

The HSE are becoming increasingly interested in work-related violence and have published a range of guidance material on the subject (see **REFS 12–18**).

Work-related travel

19.8 Far more people are killed on the roads each year whilst at work than in all other workplaces put together – it is estimated that up to a third of all road traffic accidents involve someone who is at work at the time, around 1000 road fatalities each year (**REFS 19 AND 20**). For someone driving 25 000 miles a year in a company car, it has been calculated that their probability of a fatal accident is similar to someone working in activities such as coal mining, construction and agriculture (**REF. 21**). Indeed it could be argued that an HSE Inspector has a greater chance of being killed whilst at work than most of those working in the factories that he or she visits.

Clearly many of the risks associated with road travel are outside the control of individual employers, although the Selby rail crash (caused by a vehicle getting onto the railway line after its driver dozed off at the wheel) has focused attention upon driver fatigue. Such fatigue may simply result from long hours at work (whether driving or not), whilst unrealistic work schedules might also encourage excessive speed, itself a cause of accidents. Both of these factors are very much within the control of employers.

Hours of work of drivers working in the public transport and haulage sectors are closely regulated, but many others (such as delivery drivers, sales staff and senior executives) clock up significant mileages each year without any controls at all. A macho business culture often encourages staff to work a long day and then drive a significant distance home or to the location of the next day's business – often with tragic results.

Many responsible employers are becoming more aware of the risks and addressing these cultural issues and in some cases taking other steps to reduce risks by providing defensive driving training for staff covering high mileages. Legislation now prohibits the use of hand-held mobiles whilst driving.

Whilst road transport legislation generally takes precedence over health and safety at work legislation in this area, the HSE have been taking an increasing interest in work-related driving. Their free leaflet *Driving at work* (**REF. 19**) provides excellent guidance on factors to be taken into account during risk assessments, such as:

1 Drivers and their fitness to drive

- possession of valid current driving licences;
- fitness to drive heavy vehicles;
- possible medical checks or eyesight tests;
- any medication issues.

2 Training of drivers

- awareness of company policies (e.g. re-driving hours, fatigue);
- routine safety checks (lights, tyres, etc.);
- equipment adjustments;
- loading and load distribution;
- action in case of vehicle breakdown.

3 Vehicles

- suitability for the purpose (purchased and hired vehicles);
- arrangements for maintenance and repair;
- availability of suitable tools and equipment;
- insurance and MOT certification of private vehicles.

Other aspects of driving to be taken into account during the risk-assessment process relate to the ergonomic suitability of the vehicle. Many organisations give drivers little or no input into the selection of company vehicles and as a result employees are spending many hours sitting on seats which they find uncomfortable, or are forced to adopt ergonomically undesirable driving positions.

Many cars are also used as mobile offices, with paperwork being carried out or laptop computers used within them, sometimes for extended periods – again often with little thought to ergonomic issues. Whilst vehicles are excepted from the workstation and seating requirements of the *Workplace (Health, Safety and Welfare) Regulations 1992*, work within them is still subject to the general requirements of *section 2* of *HSWA* and should therefore be included in the risk-assessment process.

The manual loading and unloading of materials into and out of vehicles is subject to the *Manual Handling Operations Regulations 1992* and the risks associated with such activities should be assessed. The handling of sample goods or exhibition stand components can often pose a significant risk to a lone member of the sales staff. Vehicles should be selected with a mind to the need to load and unload such items, the items themselves should be designed to minimise handling risks and relevant staff provided with appropriate manual handling training (see **CHAPTER 11: ASSESSMENT OF MANUAL HANDLING** for further guidance).

Whilst these ergonomic and manual handling issues are often overlooked by employers, their legal obligations in these areas are clear. However, road-traffic legislation has always tended to place responsibilities for road accidents at the doors of drivers who drove too fast or whilst tired, rather than with those who imposed unrealistic programmes on them. As a result, many directors and senior managers feel they have little or no responsibility for how their staff drive.

The DETR booklet *Revitalising Health and Safety* (**REF. 1**) committed the Government to implementing earlier recommendations by the Law Commission on a new offence of 'corporate killing' which would greatly extend the existing personal responsibilities of directors. Several years later, it is still not clear what is going to happen in this respect. However, if and when such a change is made, directors' minds will become much more focused not only on the potential for road accidents but also on the many other risks faced by their employees whilst at work.

References

19.9

1		Revitalising Health and Safety (DETR/HSC 2000)
2		*Walker v. Northumberland County Council [1995] 1 All ER 737*
3		*Sutherland v. Hatton and other appeals [2002] EWCA Civ 76*
4	INDG 406	Tackling stress. The Management Standards approach (HSE 2005)
5	MISC 714	Making the Stress Management Standards work (HSE 2005)
6	MISC 686	Working together to reduce stress at work (HSE 2005)
7	–	Real solutions, real people. A manager's guide to tackling work-related stress (HSE 2003)
8	RR 362	Farmers, farm workers and Work-Related Stress (HSE 2005)
9	CRR 435	Interventions to control stress at work in hospital staff (HSE 2002)
10	–	Managing work-related stress. A guide for managers and teachers in schools (HSE 1998)
11		The Suzy Lamplugh Trust, 14 East Sheen Avenue, London SW14 8AS. Tel: 0208 392 1839
12	HSG 133	Preventing violence to retail staff (HSE 1995)
13	HSG 100	Prevention of violence to staff in banks and building societies (HSE 1993)
14		Violence and aggression to staff in health services (HSE 1997. ISBN 0 7176 1466 2)
15		Violence in the education sector (HSE 1997. ISBN 0 7176 1293 7)
16	INDG 69	Violence at work – a guide for employers (HSE 2000)
17	HSG 229	Work-related violence. Case studies. Managing the risk in smaller businesses (HSE 2002)
18	ETIS 2	Violence to workers in broadcasting (HSE 1996)
19	INDG 382	Driving at work (HSE 2003)
20		Work-Related Road Safety Task Group: Reducing at-work road traffic incidents (HSC/DLTR 2001)
21		The tolerability of risk from nuclear power stations and DOT road casualties GB (HSE 1995)

Index

A

ACCIDENT/INCIDENT RECORDS, 4.4
ACCIDENTS:
Accident Prevention Advisory Unit, research of, 18.2
causes, 8.19
costs of accidents at work, 18.2–18.17
hazardous to health, control of substances, 9.37
investigation of, 8.17, 8.20
reports, 11.19
ratio studies, 8.18
road, 18.17
treatment records, 11.19
ACOP (APPROVED CODE OF PRACTICE AND GUIDANCE):
asbestos, 15.4, 15.6, 15.8, 15.25
carrying out of assessments, 4.4
external services, contacts with, 2.14
hazardous to health, control of substances, 9.1, 9.22, 9.37, 9.41
ionising radiations regulations, 1.20
management cycle, 8.3
model assessments, 6.1, 6.2
mothers, new or expectant, risks to:
biological agents, 3.14
chemical agents, 3.15
skin, agents absorbed through, 3.15
precautions, evaluation of, 2.3
records, assessment, 5.1
review of risk assessments, 2.7
safe systems of work, 1.23
specialised assessment techniques, 7.5
'suitable and sufficient', defined, 2.4
training requirements, 2.18
types of business, 2.5, 7.1
ACOUSTIC ENCLOSURES, 10.19
ACTIVITIES:
dangerous, 3.6
health and safety committees, 8.23
work, variations in, 4.7
ADVICE, HEALTH AND SAFETY, 2.12, 8.10
ADVISORY COMMITTEE ON DANGEROUS PATHOGENS, 3.14
AGENTS:
asbestos, 15.8
managing, 15.8
AGRICULTURE:
children and young persons, prohibitions on, 3.9
ANNUAL PLANS, HEALTH AND SAFETY, 8.5
ANTIMITOTIC DRUGS, RISKS OF, 3.15
ARRANGEMENTS:
Management Regulations, health and safety, 2.10
manual handling assessments, 11.18
ASBESTOS CONTAINING MATERIALS (ACMs)
ACOP, 15.4, 15.6, 15.8, 15.25
algorithms, 15.22
asbestos contractors, managing, 15.29
asbestosis, 15.2
assessment scope, 15.17
building work, 15.27
building workers, 15.26
categorising materials, 15.15
control of building and maintenance work, 15.27
co-operate, duty to, 15.10
costs, 15.10
delegation, 15.11
dutyholders
combinations of, 15.8
definition of, 15.7
emergency services, 15.26
guidance on, 15.5, 15.9, 15.14, 15.22, 15.31
Health and Safety Executive, 15.5, 15.9
identification surveys, 15.14
information, 15.26
landlords, 15.6
licensing, 15.4
lung cancer, 15.2
maintenance of ACMs, 15.24
work, 15.27
workers, 15.26

ASBESTOS CONTAINING MATERIALS (ACMs) – *contd*
manage asbestos, duty to, ... 1.14, 1.17, 15.1–15.30
management plan records, ... 15.28
managing agents, ... 15.8
mesothelioma, ... 15.2
monitoring ACMs, ... 15.25
multi-occupancy premises, ... 15.8
owners and occupiers, ... 15.8
records, ... 15.19, 15.28
references, ... 15.31
regulations, possible changes to, ... 19.4
removal, ... 15.24
reviews, ... 15.18, 15.30
risk management plans, ... 15.20–15.30
contents of, ... 15.23
preparation of, ... 15.21
records, ... 15.29
review, revision and implementation of, ... 15.30
risks from asbestos, ... 15.2
sampling surveys, ... 15.14
statutory requirements, ... 1.14, 15.3–15.11
survey types, ... 15.14, 15.16
full access, ... 15.14
identification, ... 15.14
presumptive, ... 15.14
sampling, ... 15.14
where asbestos might be found, ... 15.13
ASQUITH, LORD JUSTICE, ... 1.3
ASSESSING AND MANAGING RISK, ... 18.1–18.18
benefits, ... 18.7–18.9
costs of failure, ... 18.2, 18.10–18.17
examples of high profile failures, ... 18.3–18.6
ASSESSORS, RECOMMENDED:
display screen equipment, ... 12.10
fire precautions, ... 14.7
general procedures, ... 4.2
'insiders', ... 4.2
Management Regulations, ... 2.6
manual handling operations, ... 11.17
noise, ... 10.23
'outsiders', ... 4.2
planning and preparation, ... 4.2
AUDITS, HEALTH AND SAFETY, ... 8.15

B

BIOLOGICAL AGENTS, HARMFUL EXPOSURE TO:
children and young persons, ... 3.6
hazardous to health, control of substances, ... 9.4, 9.16, 9.35
mothers, new or expectant, ... 3.14
BIRD, FRANK, ... 8.19
BRITISH STANDARDS:
fire extinguishers, ... 14.15
safety of machinery, ... 5.4
BUILDING SITES:
sample PPE assessment, ... 13.11
BUILDING WORK:
asbestos, ... 15.27
BUILDING WORKERS:
asbestos, ... 15.26
BUSINESS INTERRUPTION, ... 18.16

C

CARBON MONOXIDE, PROTECTION FROM, ... 3.15
CARCINOGENS:
hazardous to health, control of substances, ... 9.16
CDM (CONSTRUCTION (DESIGN AND MANAGEMENT) REGULATIONS 1994):
ACOP, ... 18.4
contractors, manuals of, ... 5.7
co-operation, ... 2.16
co-ordination, ... 2.16
employers, host, duties of, ... 2.17
health and safety plans, .. 1.26, 17.16, 18.4
proposed changes to, ... 19.3
CHASE AUDITING SYSTEM, ... 8.15
CHECKLISTS, ASSESSMENT:
CDM Health and Safety Plans, ... 17.16
children and young people, ... 3.6
DSE workstations, ... 12.14
general issues, ... 4.8
manual handling operations: 11.20
new and expectant mothers, ... 3.18
CHEMICAL AGENTS, HARMFUL EXPOSURE TO:
children and young persons, ... 3.6
mothers, new or expectant, ... 3.15
CHEMICAL INDICATOR TUBES, ... 9.31

CHILDCARE GROUP (OUT OF SCHOOL), ILLUSTRATIVE ASSESSMENT, ... 5.4
CHILDREN AND YOUNG PERSONS:
agriculture, ... 3.9
assessing risks to, ... 3.5
biological agents, harmful exposure to, ... 3.6
chemical agents, harmful exposure to, ... 3.6
dangerous goods, ... 3.9
dangerous processes or activities, ... 3.6
dangerous workplaces or workstations, ... 3.6
defined, ... 3.3
directives, ... 3.4, 3.5, 3.6
docks, ... 3.9
explosives, carriage of, ... 3.9
information provision, ... 3.8
ionising radiation, ... 3.9
lead, ... 3.9
Management Regulations, requirements of, ... 3.4, 3.7
mines and quarries, ... 3.9
physical agents, harmful exposure to, . . 3.6
physically demanding work, excessive, ... 3.6
prohibitions on, ... 3.9
psychologically demanding work, excessive, ... 3.6
purpose of risk assessments, ... 3.7
shipbuilding and shiprepairing, ... 3.9
students, ... 3.2
work equipment, ... 3.6
CHIP (CHEMICALS (HAZARD INFORMATION AND PACKAGING FOR SUPPLY) REGULATIONS 2002), ... 3.15
CO-OPERATION:
asbestos, ... 15.10
Management Regulations requirements, ... 2.16
CO-ORDINATION:
Management Regulations requirements, ... 2.16
COCHLEA, ... 10.2
COMAH (CONTROL OF MAJOR ACCIDENT HAZARD REGULATIONS 1999):
specialised techniques, ... 7.1
statutory requirements, ... 1.19
COMMITTEES, HEALTH AND SAFETY, ... 8.21–8.24
activities, ... 8.23
communication, ... 8.9
composition, ... 8.22
consultation, ... 8.9
meetings, conduct of, ... 8.24
COMPETENCE, DEFINED, ... 2.6
COMPLIANCE SURVEYS, ... 8.16
COMPRESSED AIR WORK, EFFECT ON NEW OR EXPECTANT MOTHERS, ... 3.13
CONFINED SPACES, FIRE RISK ASSESSMENT, ... 14.8
CONSTRUCTION REGULATIONS *see* **CDM (CONSTRUCTION (DESIGN AND MANAGEMENT) REGULATIONS 1994)**
CONTRACTORS:
asbestos, ... 15.29
assessing and managing risks, ... 18.12
carrying out risk assessments, ... 4.5
Management Regulations, ... 2.17
manuals for, ... 5.7
Principal, CDM health and safety plans, ... 17.16
standards, ... 18.12
CONTROL:
carcinogens, mutagens and biological agents, ... 9.16
hazardous to health, control of substances, ... 9.13–9.15, 9.25–9.27
personal protective equipment, ... 9.14
procedures, ... 8.11
supervision, ... 8.13
systems, ... 8.11
training programmes, ... 8.12
CORPORATE KILLING, ... 19.8
COSHH (CONTROL OF SUBSTANCES HAZARDOUS TO HEALTH REGULATIONS 2002), ... 9.1–9.48
accidents, incidents and emergencies, ... 9.37
ACOP, ... 9.22, 9.37, 9.41
after the assessment, ... 9.24–9.41
armchair assessments, ... 9.44
assessment records, ... 9.22, 9.23
biological agents, ... 3.14, 9.16
biological monitoring, ... 9.35
biological problems, causing, ... 9.4

COSHH (CONTROL OF SUBSTANCES HAZARDOUS TO HEALTH REGULATIONS 2002), – *contd*

carcinogens, 9.16
changes needed, significant, 9.40
chemical agents, 3.15
chemical indicator tubes, 9.31
control, 9.10–9.16
  adequate, 9.15
  maintenance, examination and testing, 9.26, 9.27
  personal protective equipment, 9.14
  use of, 9.25
data sheet library, 9.43
definition, 9.5
direct reading instruments, 9.32
discussions, 9.19
emergencies, 9.37
entry routes, 9.3
examination of control measures, 9.26
existing control measures, 9.9
filter heads, 9.33
further tests and investigations, 9.21
harm, how substances cause, 9.2–9.4
harmful events, 9.4
health surveillance, 9.35
help, sources of, 9.20
identification of substances present, 9.9
implementation of recommendations, 9.24
incidents, 9.37
information, 9.9, 9.36
ingestion, 9.3
inhalation, 9.3
instructions, 9.36
interviews, 9.35
investigations, further, 9.21
local exhaust ventilation plant (LEV), 9.28
maintenance, 9.26, 9.27
making the assessment, 9.17–9.22
medical surveillance, 9.35
monitoring:
  biological, 9.35
  exposure, 9.30
  interpretation of results, 9.34
mothers, new or expectant, risks to:
  biological agents, 3.14
  carbon monoxide, 3.15
  chemical agents, 3.15
  drugs, 3.15
  mercury and derivatives of, 3.15
  skin, agents absorbed through, 3.15
mutagens, 9.16
observations, 9.18
organisation of, 9.8
overkill, 9.45
personal protective equipment, 9.14, 9.29
persons who will carry out the assessment, 9.7
pitfalls, 9.41–9.47
planning and preparing for the assessment, 9.6–9.9
poisoning, 9.4
preparation of assessment records, 9.22
prevention, 8.2
  control and, 9.10–9.16
  hierarchy of measures, 9.11
  risk assessment, origins of, 1.1
prevention of exposure, 9.12
previous assessments and surveys, 9.9
records of assessment, 9.22, 9.23
references, 9.48
respiratory problems, 9.4
review:
  assessments, 9.38
  recommendations, 9.24
  regular, 9.41
sample assessment records, 9.23
sampling pumps, 9.33
skin, 9.3
skin conditions, 9.4
sources of help, 9.20
tests:
  control measures, of, 9.26
  further, 9.21
training, 9.36
valid, assessments no longer, 9.39
workplace exposure limits, 9.15, 9.21
see also **HAZARDOUS SUBSTANCES,** sample model assessment

COSTS OF ACCIDENTS AT WORK, 18.2–18.6, 18.10–18.17

CUSTOMER CONFIDENCE, LOSS OF, 18.13

CYCLE, MANAGEMENT, ... 6.2, 8.3, 8.14, 8.23, 8.26
CYTOTOXIC DRUGS, RISKS OF, ... 3.15

D

DANGEROUS SUBSTANCES, ... 1.13
carriage of, ... 3.9
children and young persons, ... 3.9
DANGEROUS SUBSTANCES AND EXPLOSIVE ATMOSPHERES
assessment records, ... 14.31, 14.32
control measures, ... 14.30
definitions, ... 14.27, 14.28
records, ... 14.31–14.32
statutory requirements, ... 14.29
DATA SHEET LIBRARY, ... 9.43
DECIBELS, ... 10.4
DEFINITIONS:
asbestos duty holders, ... 15.7
children and young persons, ... 3.3
competence, ... 2.6
construction work, ... 16.16
dangerous substances, ... 14.27
display screen equipment, ... 12.2
dynamic risk assessment, ... 1.24
explosive atmosphere, ... 14.28
hazards, ... 2.2, 9.5
manual handling operations, ... 11.2
mothers, new or expectant, ... 3.10
PPE (personal protective equipment), ... 13.2
practicable and absolute requirements, ... 1.4
'reasonably practicable', ... 1.3
risks, ... 2.2
safe system of work, ... 17.2
'suitable and sufficient', ... 2.4
task procedures, ... 17.4
task safety instructions, ... 17.3
work at height, ... 16.2
DETR (DEPARTMENT OF THE ENVIRONMENT, TRANSPORT AND THE REGIONS), ... 18.1
DIRECT READING INSTRUMENTS, ... 9.32
DIRECTORS LIABILITY, ... 19.8
DISCUSSIONS, IMPORTANCE OF:
carrying out risk assessments, ... 4.11
DISCUSSIONS, IMPORTANCE OF: – *contd*
display screen equipment, ... 12.13
fire safety, ... 14.19
hazardous to health, control of substances, ... 9.19
manual handling operations, ... 11.22
DISPLAY SCREEN EQUIPMENT (DSE):
assessment approach, ... 12.11
assessment records, ... 12.14
definitions, ... 12.2
following assessment:
re-assessments, ... 12.21
recommendations, review and implementation, ... 12.20
review of assessment, ... 12.21
homeworkers, ... 12.16
hot desking, ... 12.17
HSE guidance, ... 12.14, 12.15, 12.22
laptops, ... 12.18
making the assessments:
discussions, ... 12.13
observations, ... 12.12
records, ... 12.14
mobile phones, ... 12.18
new or expectant mothers, effect on, ... 3.13
personal organisers, ... 12.18
planning and preparation:
assessors, ... 12.10
organisation of assessments, ... 12.11
pointing devices, ... 12.19
portable computers, ... 12.18
self-assessment checklist, ... 12.14
shared workstations, ... 12.17
statutory requirements, ... 1.10, 12.2–12.9
teleworkers, ... 12.16
training, ... 12.8, 12.16
'users', identification of, ... 12.11
DISRUPTION OF WORK, ... 18.9
DIVING WORK, EFFECT ON NEW OR EXPECTANT MOTHERS, . . 3.13
DOCKS:
children and young persons, prohibitions on, ... 3.9
DOCUMENTS, GATHERING OF, ... 4.4

DOMINO MODEL, LOSS CAUSATION, 8.19
DOSEMETERS, PERSONAL NOISE, 10.6
DRIVING, WORK-RELATED, 19.8
DRUGS, RISK TO NEW OR EXPECTANT MOTHERS, 3.15
DES *see* **DISPLAY SCREEN EQUIPMENT (DSE)**
DYNAMIC RISK ASSESSMENTS, 1.24, 17.8, 17.9
CDM Health and Safety Plans, 17.16
working at height, 16.14

E

EAR DEFENDERS, 10.21
EAR MUFFS, 10.21
EAR PLUGS, 10.21
EAR PROTECTION, 10.12, 10.21, 10.28
zones, 10.11, 10.26, 10.28, 10.29, 10.30
EARDRUM, 10.2, 10.3
ELECTRICITY AT WORK:
statutory provisions, 5.4
ELECTROMAGNETIC RADIATION, EFFECT ON NEW OR EXPECTANT MOTHERS, 3.13
EMERGENCY:
assessment and management by emergency services, 17.9
hazardous to health, control of substances, 9.37
services, ACM information, 15.26
violence at work, 19.7
EMPLOYEES:
absence of, 18.14
assessing and managing risk, 18.1
capabilities, 2.18
carrying out risk assessments, 4.5
consultation of, 8.21–8.24, 18.7
guides for, 5.6
handbooks for, 5.6, 5.7
industrial action, 18.16
information provision, 2.15, 2.17, 10.15, 10.29
Management Regulations, information provision, 2.15
PPE, statutory duties under, 13.5
safe, healthy and well-motivated, 18.7
EMPLOYEES: – *contd*
'safe person' model, 17.9
students as, 3.2
visiting, 2.17
EMPLOYERS:
employers' liability insurance, 18.11
persons working in undertakings of, 2.17
self regulation, 19.1
EQUIPMENT REQUIREMENTS:
method statements, 17.17
working at height, 16.4, 16.5, 16.6, 16.7
ESTATE AGENTS' OFFICES, ILLUSTRATIVE ASSESSMENTS, 5.4
EVENT TREE ANALYSIS, 7.5
EXHAUST VENTILATION PLANT, LOCAL, 9.28
EXPLOSIVES:
carriage of, 3.9
children and young persons, prohibitions on, 3.9
EXPLOSIVE ATMOSPHERES see DANGEROUS SUBSTANCES AND EXPLOSIVE ATMOSPHERES
EXTERNAL ACTIVITIES, SAMPLE ASSESSMENT RECORD, 5.4
EXTERNAL SERVICES, CONTACTS WITH, 2.14
EYE TESTS, 12.7
EYESIGHT PROTECTION:
display screen legislation, 12.1–12.9

F

5 STAR AUDIT, 8.15
FAMILY UNDERTAKING, 3.8
FAULT TREE ANALYSIS, 7.5
FILTER HEADS, 9.33
FINES, 18.4–18.6, 18.10
***FIRE SAFETY: AN EMPLOYER'S GUIDE* (HOME OFFICE PUBLICATION),** 14.5, 14.9
FIRE SAFETY PRECAUTIONS:
alarms, 14.12, 14.14, 14.24
call points, 14.17
assembly points, designated, 14.14
assessment records, 6.3, 14.22
detection of fire, 14.11, 14.24
emergency lighting, 14.24
equipment, fire fighting, 14.15, 14.17, 14.24

FIRE SAFETY PRECAUTIONS: – *contd*
escape routes, ... 14.13, 14.17
evacuation procedures, ... 14.14, 14.17, 14.24
Fire Certificates, ... 14.9
following assessment:
maintenance, ... 14.24
recommendations, review and implementation, ... 14.23
review of assessment, ... 14.24
guidance, ... 14.5
making the assessment:
discussions, ... 14.19
observations, ... 14.18
records, ... 14.20, 14.21
planning and preparation:
assessors, ... 14.7
information gathering, ... 14.9
organisation of assessment, ... 14.8
plans and drawings, ... 14.9
prevention measures, ... 14.16, 14.17
records of assessment, ... 6.3, 14.22
reference material, ... 14.9
Regulatory Reform (Fire Safety) Order 2005, ... 14.2–14.6
risk assessment factors, ... 14.10–14.17
detection of fire, ... 14.11, 14.24
escape routes, ... 14.13
evacuation procedures, . . . 14.14, 14.17, 14.24
fire fighting, ... 14.15
prevention, ... 14.16
signs, ... 14.17
warnings, ... 14.12
sample assessment records, ... 6.3, 14.22
signs, location of, ... 14.17
smoke alarms, ... 14.11
statutory requirements:
definitions and duties, ... 14.3
enforcement, ... 14.2
fire safety duties, ... 14.4
guidance, ... 14.5
risk assessment under the Order, . . 14.6
FMEA (FAILURE MODE AND EFFECTS ANALYSIS), ... 7.5
FRAGILE SURFACES, ... 16.2

G

GENERIC RISK ASSESSMENTS, . . 17.8
GOODS, DANGEROUS, CARRIAGE OF, ... 3.9

H

HANDBOOKS, SAFETY, ... 4.4
HATFIELD RAIL CRASH, ... 18.6
HAZARDOUS SUBSTANCES:
sample model assessment, ... 6.3
see also **COSHH (CONTROL OF SUBSTANCES HAZARDOUS TO HEALTH REGULATIONS 2002)**
HAZARDS, Management Regulations, ... 2.2
HAZOP (HAZARD AND OPERATING STUDIES), ... 7.5, 7.6
HEALTH AND SAFETY:
annual plans, ... 8.5
audits, ... 8.15
committees, ... 8.21–8.24
inspections, ... 8.14
reports of, ... 4.4
Management Regulations:
arrangements, ... 2.10
assistance, ... 2.12
surveillance, ... 2.11
policy statements, ... 8.4
see also **CDM (CONSTRUCTION (DESIGN AND MANAGEMENT) REGULATIONS 1994), health and safety plans**
HEALTH AND SAFETY EXECUTIVE
see **HSE (HEALTH AND SAFETY EXECUTIVE)**
HEALTH SURVEILLANCE:
hazardous to health, control of substances, ... 9.35
Management Regulations, ... 2.11
noise, ... 10.32
HEARING, DAMAGE TO, ... 10.2, 10.3
reduction of risk, ... 10.10
HEATHROW TUNNEL COLLAPSE, ... 18.5
HEIGHT, WORK AT:
competence, ... 16.2
construction work, ... 16.2, 16.12
danger areas, ... 16.2
definitions, ... 16.2
duties of person at work, ... 16.2
dynamic risk assessments, ... 16.14
equipment, ... 16.2, 16.4, 16.5, 16.6
falling objects, ... 16.2
fragile surfaces, ... 16.2

HEIGHT, WORK AT: – *Contd*
inspections, 16.2, 16.7
ladders, 16.2, 16.8
making assessment, 16.10
organisation and planning, 16.2, 16.9
permits to work, 16.13
recording assessment, 16.11
regulations, 16.2
risk avoidance, 16.2, 16.3
sample assessment record for, 16.12
HEINRICH (ACCIDENT RESEARCHER), 8.18, 8.19
HOMEWORKERS, 12.16
HOT DESKING, 12.17
HOURS OF WORK, 19.8
HSE (HEALTH AND SAFETY EXECUTIVE):
display screen equipment, 12.1
Employment Medical Advisory Service, 3.11
guidance:
asbestos, 15.5, 15.9, 15.14, 15.22, 15.31
carrying out of assessments, 4.4
CDM health and safety plans, 17.16
display screen equipment, 12.14, 12.15, 12.22
manual handling operations, 11.19, 11.24, 11.27
mothers, new or expectant, 3.13, 3.14, 3.15
noise control, 10.7, 10.16, 10.20, 10.21
mothers, new or expectant, risks to:
biological agents, 3.14
compressed air, 3.13
mercury and derivatives of, 3.15
temperature, extreme, 3.13
publications:
Carbon monoxide, 3.15
Controlling Noise at Work, 10.7, 10.16, 10.20, 10.21, 10.28, 10.32
Driving at Work, 19.8
Essentials of Health and Safety at Work, 4.4
Five steps to risk assessment, 2.6, 5.3
Getting to grips with manual handling, 11.27
Health and safety in motor vehicle repair, 5.4
Maintaining portable electrical equipment in offices and other low–risk environments, 5.4

HSE (HEALTH AND SAFETY EXECUTIVE): – *contd*
Management of health and safety at work, 2.1
Managing asbestos in premises, 15.19, 15.22, 15.24, 15.29
Manual handling assessment charts, 11.19
Manual handling: solutions you can handle, 11.10
Methods for the Determination of Hazardous Substances, 9.34
Need help on health and safety?, 2.6
New and expectant mothers at work, 3.12
Office wise, 5.4
Quantified risk assessment, 7.5
Safe use of pesticides, 3.15
Sound solutions, 10.16
Workplace exposure limits, 3.15, 9.15
Workplace Transport Safety, 5.4
Young people at work. A Guide for Employers, 3.3, 3.8
reputation with, 18.8
self-regulation, 19.1
stress, 19.6
visits from, 18.8
website, 4.4
HSWA (HEALTH AND SAFETY AT WORK ETC ACT 1974):
guidance information, 4.4
key requirements, 1.2
noise, information on, 10.16
policy statements, 8.4
'reasonably practicable', burden of proof, 1.3
safe systems of work, 1.23
training requirements, 2.18

I

ILL HEALTH RECORDS, 11.19
INCIDENT INVESTIGATION, 8.17, 8.20, 9.37
INDUSTRIAL ACTION, 18.16
INFORMATION (GATHERING AND PROVISION):
asbestos, 15.26
children and young persons, 3.8
employees, 2.15, 2.17, 10.15, 10.29
employers, duties imposed on, 8.10
fire risk assessment, 14.9
hazardous to health, control of substances, 9.9, 9.36

INFORMATION (GATHERING AND PROVISION): – *Contd*
manual handling assessments, 11.19
noise assessments, 10.14, 10.25, 10.27
trade, 4.4, 11.19
workstations, DSE, 12.9
INHALATION AND INGESTION, 9.3
INSPECTIONS, HEALTH AND SAFETY, 8.14
INSTITUTION OF OCCUPATIONAL SAFETY AND HEALTH (IOSH), 2.6, 7.2
INSURANCE
costs of accidents at work, 18.2, 18.11, 18.15
employers' liability insurance, 18.11
INTEGRATING SOUND LEVELS METERS, 10.6
INTEGRATION OF RISK ASSESSMENT REQUIREMENTS, 19.2
INTERNATIONAL SAFETY RATING SYSTEM (ISRS), 8.15
INVESTIGATIONS
accident and incident, 8.17, 8.20
further requirements, 4.13
hazardous to health, control of substances, 9.21
IONISING RADIATION
children and young persons, prohibitions on, 3.9
mothers, new and expectant, 3.13
statutory requirements, 1.20
IOSH (INSTITUTION OF OCCUPATIONAL SAFETY AND HEALTH), 2.6, 7.2
ISRS (INTERNATIONAL SAFETY RATING SYSTEM), 8.15

K

KEY PERSONNEL IDENTIFICATION
method statements, 17.17

L

LADDERS, 16.8
LANDLORDS
asbestos, 15.6
LANARKSHIRE GAS EXPLOSION, 18.4
LARGE BUSINESSES
accident investigation, 8.20
health and safety committees, 8.21
Management Regulations, 2.5
specialised techniques, 7.1
LEAD AT WORK
children and young persons, prohibitions on, 3.9
mothers, new or expectant, 3.15
statutory requirements, 1.16
LEGAL COSTS
assessing and managing risk, 18.10
LICENCES
asbestos, 15.4
LIFTING EQUIPMENT AND OPERATIONS
sample model assessment, 6.3
statutory provisions, 1.23, 3.4
see also **MANUAL HANDLING OPERATIONS**
LIQUIDS, HIGHLY FLAMMABLE, 14.5
LOADING OPERATIONS, 19.8
LOCAL EXHAUST VENTILATION (LEV) PLANT, 9.28
LOLER (LIFTING OPERATIONS AND LIFTING EQUIPMENT REGULATIONS 1998), 1.23, 3.4
LOSS CAUSATION, DOMINO MODEL, 8.19
LUNG CANCER, 15.2

M

MACHINERY SUPPLY, SAFETY OF
statutory requirements, 1.18
MAINTENANCE
asbestos, 15.24, 15.26, 15.27
hazardous to health, control of substances, 9.26–9.29
hearing protection, 10.14
PPE, 13.4
work, 15.27
workers, 15.26
MAJOR ACCIDENT HAZARDS, CONTROL OF *see* **COMAH (CONTROL OF MAJOR ACCIDENT HAZARD REGULATIONS 1999)**

MAKING THE ASSESSMENT
discussions, carrying out
assessments, ... 4.11
fire precautions
discussions, ... 14.19
observations, ... 14.18
records, ... 14.20, 14.21
further investigations, ... 4.13
hazardous to health, control of
substances, ... 9.17–9.22
manual handling operations, ...
11.20–11.23
noise
conclusions, drawing, ... 10.29
information required, ... 10.28
purpose, ... 10.26
records, ... 10.30
notes, ... 4.14
observations, carrying out risk
assessments, ... 4.10
records, ... 4.15
tests, ... 4.12
work at height, ... 16.10
MANAGEMENT CYCLE, ... 6.2, 8.3,
8.14, 8.23, 8.26, 18.9
MANAGEMENT OF RISK *see*
ASSESSING AND MANAGING RISK
MANAGEMENT REGULATIONS, ... 2.1–2.19
arrangements, health and safety, ... 2.10
assessors, recommended, ... 2.6
assistance, health and safety, ... 2.12
capabilities, ... 2.18
children and young persons, requirements
relating to, ... 3.4, 3.8
co-operation, ... 2.16
co-ordination, ... 2.16
cycle, management, ... 8.3
danger, procedures for, ... 2.13
employees, information
provision for, ... 2.15
external services, contacts with, ... 2.14
hazards, ... 2.2
health surveillance, ... 2.11
host employers' undertakings, ... 2.17
large businesses, ... 2.5
lead regulations, ... 1.16
medium sized businesses, ... 2.5
mothers, new or expectant, ... 3.11
precautions, evaluation of, ... 2.3
MANAGEMENT REGULATIONS, – *Contd*
prevention principles, ... 2.9, 8.2
purpose, ... 1.5
records, assessment, ... 5.1
review of risk assessments, ... 2.7
risks, ... 2.2
self-employed persons,
undertakings of, ... 2.17
small businesses, ... 2.5
statutory requirements, ... 1.5, 2.1
'suitable and sufficient', meaning, ... 2.4
training, ... 2.18
MANAGING AGENTS, ... 15.8
MANUAL HANDLING OPERATIONS, ... 11.2–11.27
age, ... 11.7
assessment records, ... 11.24
avoiding or reducing risks
automation, ... 11.10
capability of individual, ... 11.15
handling, elimination of, ... 11.9
load, reduction of, ... 11.11
load-related measures, ... 11.13
mechanisation, ... 11.10
task-related measures, ... 11.12
work environment, improving, .. 11.14
body movement, unsatisfactory, ... 11.4
carrying, ... 11.3
excessive movement of loads, ... 11.4
excessive pushing or pulling, ... 11.4
experience, ... 11.7
fitness programmes, ... 11.15
floors
improving, ... 11.14
uneven, slippery or unstable, ... 11.6
variation in levels, ... 11.6
following assessment
recommendations, review and
implementation, ... 11.25
records, ... 11.24
review of assessment, ... 11.26
training, ... 11.27
gender, ... 11.7
handling when seated, ... 11.3
humidity, ... 11.6
individual, capability of, ... 11.7
accommodating, ... 11.15
lifting, ... 11.3
lighting, poor, ... 11.6
measures for, ... 11.14

MANUAL HANDLING OPERATIONS, – *Contd*
loads
bulk, 11.5
excessive movement, 11.4
grasp, 11.5, 11.13
heaviness, excessive, 11.5
holding or manipulating at distance from trunk, 11.4
reducing, 11.11, 11.13
sharp, 11.5, 11.13
stability, 11.5, 11.13
temperature, 11.5
unwieldy, 11.5
lowering, 11.3
making the assessment, 11.20
mechanical assistance, 11.12
medical screening, 11.15
mothers, new or expectant, 3.13
physical effort, frequent or prolonged, 11.4
physique, 11.7
planning and preparation
arrangement of assessments, 11.18
assessors, recommended, 11.17
information gathering, 11.19
posture, unsatisfactory, 11.4
previous injury, 11.7
process, rate of work imposed by, 11.4
pulling, 11.3
excessive, 11.4
pushing, 11.3
excessive, 11.4
records of assessment, 11.24
rest and recovery, insufficient, 11.4
restrictions, long or short term, 11.15
risk of injury
capability of individual, 11.7
guidelines, 11.3
load related factors, 11.5
PPE, use of, 11.8
task-related factors, 11.4
uniforms, 11.8
working environment, factors relating to, 11.6
space constraints, 11.6
measures for, 11.14
stature, 11.7
statutory requirements, 1.9, 11.1, 11.2
stooping, 11.4

MANUAL HANDLING OPERATIONS, – *Contd*
sudden movement, risk of, 11.4
task layout and design, 11.12
team handling, 11.12
temperature, 11.5, 11.6
measures for, 11.14
training, 11.15
trunk, twisting, 11.4
vehicles and manual handling, 19.8
ventilation problems, 11.6
wind, gusts of, 11.6
measures for, 11.14
work routines, 11.12
work surfaces, variation in levels, 11.6
work surfaces, variations in levels, improving, 11.15

MANUFACTURERS
guidance from, 4.4
noise information, duty to provide, 10.20

MAXIMUM EXPOSURE LIMIT (MEL) *see* **WORKPLACE EXPOSURE LIMITS**

MEDIUM-SIZED BUSINESSES
health and safety committees, 8.21
Management Regulations, 2.5
specialised techniques, 7.1

MEETINGS
health and safety committees, 8.24
management, 8.25

MERCURY, RISKS FROM, 3.15

MESOTHELIOMA, 15.2

METHOD STATEMENTS, 1.27, 17.17

MINES AND QUARRIES
children and young persons, prohibitions on, 3.9

MOBILE OFFICES, CARS USED AS, 18.11

MOBILE PHONES, 12.18

MODEL ASSESSMENTS
adaptation, 6.2
implementation, 6.2
samples, 6.3

MONITORING OF PRECAUTIONS
accident and incident investigations, 8.17, 8.20
accident ratio studies, 8.18
asbestos, 15.25
audits, 8.15
causes of accidents, 8.19

MONITORING OF PRECAUTIONS – *Contd*
committees, 8.22, 8.23
compliance surveys, 8.16
hazardous to health, control of substances, 9.30, 9.34, 9.35
inspections, 8.14
investigations, purposes of, 8.20
safety observations, 8.16
safety sampling, 8.16
safety surveys, 8.16
safety tours, 8.16

MOTHERS, NEW OR EXPECTANT
biological agents, 3.14
chemical agents, 3.15
definitions, 3.10
employment rights legislation, 3.11
equal pay legislation, 3.11
Management Regulations, requirements of, 3.11
physical agents, 3.13
practical issues, 3.18
pregnancy, aspects of, 3.17
risks to, 3.12
sex discrimination legislation, 3.11
working conditions, 3.16

MOTOR TRADE, SAMPLE MODEL ASSESSMENT, 6.3

MOTOR VEHICLE REPAIR WORKSHOP, ILLUSTRATIVE ASSESSMENT, 5.4

N

NEWSPAPER PUBLISHER, ILLUSTRATIVE ASSESSMENT, 5.4

NOISE ASSESSMENT
acoustic enclosures, 10.19
assessment of exposure, 10.9
assessment records, 10.30
damage to hearing, 10.2, 10.3
dosemeters, 10.6
ear muffs, 10.21
ear plugs, 10.21
ear protection, 10.11, 10.21, 10.28
  costs, 10.21
  zones, 10.11, 10.29
employees, provision of information to, 10.14

NOISE ASSESSMENT – *Contd*
equipment
  maintenance and use of, 10.12
  specification, 10.20
exposure action values, 10.5
exposure limit values, 10.5
exposure reduction, 10.16–10.21, 10.28
  ear protection, 10.21
  equipment specification, 10.20
  generation of noise, 10.18
  statutory requirements, 10.7–10.15
  transmission of noise, 10.19
following
  possible health surveillance, 10.32
  recommendations, review and implementation, 10.31
  review of assessment, 10.33
gases, vibration in, 10.18
hearing cells, 10.2
hearing damage, 10.2, 10.3
  reduction of risk, 10.10
information, 10.14, 10.25, 10.27
maintenance
  efficient, 10.18
  equipment, 10.12
manufactures, 10.20
measurement, 10.4–10.6
  decibel scale, 10.4
  dosemeters, personal noise, 10.6
  instruments, 10.6
  personal noise exposure, 10.5
  sound level meters, 10.6
mothers, new or expectant, 3.13
personal preferences, 10.21
planning and preparation
  assessors, 10.23
  information gathering, 10.25
  organisation, 10.24
records of assessment, 10.30
reduction of noise exposure, 10.10, 10.11, 10.16–10.21, 10.29
reduction of noise generation, 10.18
reduction of noise transmission, 10.19
semi-inserts, 10.21
separation measures, 10.19
sound absorbent materials, 10.19
statutory requirements
  assessment of risk, 10.9
  elimination or control of exposure, 10.10

NOISE ASSESSMENT – *Contd*
employees, provision of
information to, 10.14
equipment, maintenance and
use, 10.12
exposure action values, 10.5
exposure limit values, 10.5
health surveillance, 10.13
hearing protection, 10.11
manufacturers, duties of, 10.20
vibrating surfaces, 10.18
vibration in gases, 10.18
work limitations, 10.21
NOTE TAKING
carrying out risk assessments, 4.14
manual handling operations, 11.23

O

OBSERVATION OF RISKS
display screen equipment, 12.12
fire safety assessments, 14.18
general procedures, 4.10
manual handling operations, 11.21
safety, 8.16, 9.18
OCCUPATIONAL EXPOSURE STANDARD (OES) *see* **WORKPLACE EXPOSURE LIMITS**
OFFICES, SAMPLE ASSESSMENTS
estate agents, 5.4
models, 6.3
ORGANISATION OF ASSESSMENTS
display screen equipment, 12.11
fire risk, 14.8
general procedures, 4.3
noise, 10.24
precautions, implementation of
advice, 8.10
communication, 8.9
consultation, 8.9
information provision, 8.10
responsibilities, 8.8
work at heights, 16.2, 16.9
ORIGINS OF RISK ASSESSMENT, 1.2–1.5

P

PENALTIES, 18.2, 18.4–18.6, 18.10, 18.16
PERFORMANCE STANDARDS, DEVELOPMENT OF, 8.6
PERMITS TO WORK
designation of permit situations, 17.11
forms, 17.13
generally, 1.25
issuers of, 17.14
authorisation of, 17.15
sequence of systems, 17.12
work at height, 16.13
PERSONAL ORGANISERS, 12.18
PERSONAL PROTECTIVE EQUIPMENT *see* **PROTECTIVE EQUIPMENT, PERSONAL (PPE)**
PESTICIDES, CONTROL OF, 3.15
PETROLEUM GAS, LIQUEFIED, 14.5
PHYSICAL AGENTS, HARMFUL EXPOSURE TO
children and young persons, 3.6
mothers, new or expectant, 3.13
PHYSICALLY DEMANDING WORK, EXCESSIVE, 3.6
PIPER ALPHA, 18.3
PLANNING AND PREPARATION
asbestos management, 15.20–15.21, 15.23
carrying out assessments, 4.2
display screen workstations
assessors, 12.10
organisation of assessments, 12.11
documents, gathering, 4.4
fire risk assessment
assessors, 14.7
information gathering, 14.9
organisation, 14.8
hazardous to health, control of
substances, 9.6–9.9
issues, 4.6
manual handling operations
arrangement of assessments, 11.18
assessors, recommended, 11.17
information gathering, 11.19
noise assessments
assessors, 10.23
information gathering, 10.25
organisation, 10.24
organisation of assessments, 4.3
persons at risk, 4.5
precautions, implementation of
annual health and safety plans, 8.5
health and safety policy
statement, 8.4
performance standards,
development of, 8.6

PLANNING AND PREPARATION – *Contd*
risk assessments, 8.7
vulnerable persons, 4.5
work activities, variations in, 4.7
POINTING DEVICES, 12.19
POISONING, 9.4
POLICY STATEMENTS, HEALTH AND SAFETY, 8.4
PORTABLE COMPUTERS, 12.18
POST-ASSESSMENT ACTIVITIES
carrying out assessments
action plan, implementation of, 4.18
assessment review, 4.20
follow-up, 4.19
recommendations, review, 4.17
display screen equipment
assessment review, 12.21
re-assessments, 12.21
recommendations, review and implementation, 12.20
fire safety
assessment review, 14.25
precautions, maintenance of, 14.24
recommendations, review and implementation, 14.23
manual handling
assessment review, 11.26
recommendations, review and implementation, 11.25
training, 11.27
noise assessment
assessment review, 10.33
recommendations, review and implementation, 10.31
PPE *see* **PROTECTIVE EQUIPMENT, PERSONAL (PPE)**
PRACTICABLE AND ABSOLUTE REQUIREMENTS, MEANING, 1.4
PRECAUTIONS
fire, 1.12
implementation, 8.1–8.26
control, 8.11–8.13
management cycle, 8.3, 8.14, 8.26
monitoring, 8.14–8.20
organisation, 8.8–8.10
planning, 8.4–8.7
principles of prevention, 8.2
review, 8.21–8.26
PRECAUTIONS – *Contd*
Management Regulations, evaluation under, 2.3, 8.2
'reasonably practicable', 19.1
PREGNANT MOTHERS *see* **MOTHERS, NEW OR EXPECTANT**
PRESSURE, EXTREME: EFFECT ON NEW OR EXPECTANT MOTHERS, 3.13
PREVENTION PRINCIPLES, 8.2
PROCEDURES FOR RISK ASSESSMENT
carrying out assessments, 4.2–4.20
checklist of possible risks, 4.8
following, 4.16–4.20
making, 4.9–4.15
planning and preparation, 4.2–4.7
PPE, 13.7
checklist of risks, 4.8
danger, serious and imminent, 2.13
following assessment
action plan, implementation of, 4.18
recommendation follow, up, 4.19
review, 4.17, 4.20
investigation, 8.20
making the assessment
display screen equipment, 12.12–12.14
fire risk, 14.18–14.21
generally, 4.10–4.15
noise, 10.26–10.30
Management Regulations, 2.13
operating procedures, 4.4, 5.8, 11.19
planning and preparation
assessors, recommended, 4.2
documents, gathering, 4.4
issues, 4.6
organisation of assessments, 4.3
persons affected, 4.5
work activities, variations in, 4.7
precautions, implementation of, 8.11
PROCESSES, DANGEROUS, 3.6
PROHIBITION NOTICES, 18.16
PROSECUTIONS, 18.10
PROTECTIVE EQUIPMENT, PERSONAL (PPE):
asbestos, 15.2
assessment records, 13.8–13.11
body protection, 13.6
ear protectors, 10.11

PROTECTIVE EQUIPMENT, PERSONAL (PPE): – *Contd*
foot protection, 13.6
hand and arm protection, 13.6
hazardous to health, control of substances, 9.14, 9.29
head protection, 13.6
hearing protection, 10.11
practical considerations
carrying out assessments, 13.7
records of assessment, 13.8–13.11
specific requirements, 13.6
statutory requirements
accommodation, 13.4
assessment of, 13.3
correct use, 13.5
definitions, 13.2
generally, 1.11, 13.1
maintenance, 13.4
replacement, 13.4
PSYCHOLOGICALLY DEMANDING WORK, EXCESSIVE, 3.6
PTW *see* **PERMITS TO WORK**
PUWER (PROVISION AND USE OF WORK EQUIPMENT REGULATIONS 1998):
illustrative assessments, 5.4
prevention principles, 8.2

Q

QRA (QUANTIFIED RISK ASSESSMENT), 7.5
QUARRIES
children and young persons, prohibitions on, 3.9

R

RADIATIONS, IONISING *see* **IONISING RADIATION**
RADIOACTIVE ANTI-STATIC DEVICES, SAMPLE MODEL ASSESSMENT, 6.3
RAILWAYS
Hatfield rail crash, 18.6
RATIO STUDIES, 8.18
'REASONABLY PRACTICABLE'
and COSHH, 8.2
meaning, 1.3
RECOMMENDATIONS, REVIEW AND IMPLEMENTATION
asbestos, 15.30
assessing and managing risk, 18.1
display screen equipment, 12.20
fire risk assessment, 14.23
hazardous to health, control of substances, 9.24–9.41
manual handling operations, 11.25
noise assessment, 10.31
precautions, 18.1
RECORDS, ASSESSMENT, 5.1–5.8
alternative formats, 5.5–5.8
asbestos, 15.19, 15.28
contents, 5.2
contractors' manuals, 5.6
display screen equipment, 12.14
DSEAR, 14.31, 14.32
employees, guides or handbooks for, 5.6, 5.7
fire safety precautions, 14.21, 14.22
format, sample, 4.15
hazardous to health, control of substances, 9.6, 9.22, 9.23, 9.47
headings, 4.15
illustrative, 5.4
manual handling operations, 11.24
noise, 10.30
PPE, 13.8–13.11
preparation, 4.15
recommendations, identification, 4.15
sample format, 5.3
standard operating procedures, 4.4, 5.8
work at height, 16.11–16.13
RECTIFICATION COSTS, 18.15
REFERENCE BOOKS, 4.4
REPAIRS, 18.15
REPLACEMENTS, 18.15
REPORTS, INVESTIGATION, 8.20
REPUTATION, 18.8
REQUIREMENTS, PRACTICABLE AND ABSOLUTE, 1.4
RESPIRATORY PROBLEMS, 9.4
REVIEW OF RISK ASSESSMENTS
asbestos, 15.18, 15.30
display screen equipment, 12.21
fire precautions, 14.25
hazardous to health, control of substances, 9.24, 9.38, 9.41
Management Regulations, 2.7
manual handling operations, 11.26

REVIEW OF RISK ASSESSMENTS – *Contd*
noise, 10.31, 10.33
precautions, implementation, 8.21–8.26
procedures, 4.17, 4.20
see also **RECOMMENDATIONS, REVIEW AND IMPLEMENTATION**
RIDDOR (REPORTING OF INJURIES, DISEASES, AND DANGEROUS OCCURRENCES REGULATIONS 1995), 8.20
RISK RATING MATRICES
advantages, 7.3
disadvantages, 7.4
examples, 7.2
RISKS
Management Regulations, 2.2
purpose of assessments, 3.7
see also **ASSESSING AND MANAGING RISK**
ROAD ACCIDENTS, 19.8
ROBENS COMMITTEE (1972), 1.2, 1.3, 18.1

S

'SAFE PERSON' CONCEPT, 17.9, 17.10
SAFE SYSTEMS OF WORK
defined, 17.2
factors in establishing, 17.5
implementation, 17.7
requirements, 1.23
task procedures, 17.4
development of, 17.6
task safety instructions, 17.3
violence, 19.7
SAFETY & HEALTH PRACTITIONER, 7.2
SAFETY HANDBOOKS, 4.4
SAFETY OBSERVATIONS, 8.16
SAFETY SAMPLING, 8.16
SAFETY SURVEYS, 8.16
SAFETY TOURS, 8.16
SAMPLING
asbestos, 15.16
hazardous to health, control of substances, 9.31–9.33
pumps, 9.33
SAMPLING – *Contd*
safety, 8.16
surveys, 15.16
SELF-EMPLOYED PERSONS
Management Regulations, 2.17
SELF REGULATION, 18.1
SERVICES, TO AN OFFICE
sample model assessment, 6.3
SHARED WORKSTATIONS, 12.17
SHIPBUILDING/SHIPREPAIRING
children and young persons, prohibitions on, 3.9
SHOCKS, PHYSICAL: EFFECT ON NEW OR EXPECTANT MOTHERS, 3.13
SITE REQUIREMENTS
method statements, 17.17
SKIN
agents absorbed through, dangers of, 3.15
hazardous to health, control of substances, 9.3, 9.4
SMALL BUSINESSES
Management Regulations, 2.5
specialised techniques, 7.1
SMOKE ALARMS, 14.11
SOUND ABSORBENT MATERIALS, APPLYING, 10.19
SOUND LEVEL METERS, 10.6
SPECIALISED TECHNIQUES, 7.1–7.6
event tree analysis, 7.5
fault tree analysis, 7.5
FMEA (failure mode and effects analysis), 7.5
HAZOP (hazard and operating studies), 7.5, 7.6
risk rating matrices, 7.2–7.4
type of business, 7.1
STANDARDS
asbestos, 15.11
contractors, 18.12
performance, development of, 8.6
sub-contractors, 18.12
STRESS, 3.13, 19.6
causes of, 19.6
costs of, 19.6
HSE booklet on, 19.6
STUDENTS, 3.2
SUB-CONTRACTORS, STANDARDS OF, 18.12

'SUITABLE AND SUFFICIENT', DEFINED, 2.4
SUPERMARKETS, ILLUSTRATIVE ASSESSMENT, 5.4
SUPERVISORS, RESPONSIBILITIES OF, 8.13
SUPPLIERS, GUIDANCE FROM, 4.4
SUPPLY-CHAIN PRESSURE, 18.12
SURVEILLANCE, HEALTH AND SAFETY *see* **HEALTH SURVEILLANCE**
SURVEYS
asbestos, 15.14, 15.16
hazardous to health, control of substances, 9.9
safety and compliance, 8.16

T

TASK PROCEDURES, 17.4
development of, 17.6
TASK SAFETY INSTRUCTIONS, 17.3
TECHNIQUE, INVESTIGATION, 8.20
TECHNIQUES, SPECIALISED *see* **SPECIALISED TECHNIQUES**
TELEWORKERS, 12.16
TEMPERATURE, EXTREME: EFFECT ON NEW OR EXPECTANT MOTHERS, 3.13
TESTS
hazardous to health, control of substances, 9.21, 9.26
safety precautions, 4.12
TIMBER-FRAMED STRUCTURES ON BUILDING SITES, ERECTION OF, 13.11
TINNITUS, 10.3
TOURS, SAFETY, 8.16
TRADE INFORMATION, 4.4, 11.19
TRAINING PROGRAMMES
carrying out of assessments, 4.4
precautions, control of, 8.12
TRAINING REQUIREMENTS
display screen equipment, 12.8, 12.16
hazardous to health, control of substances, 9.36–9.37
Management Regulations, 2.18
TRAINING REQUIREMENTS – *Contd*
manual handling operations, 11.15, 11.27
method statements, 16.17
TRAVEL, 19.8
cars used as mobile offices, 19.8
hours of work, 19.8
loading operations, 19.8
road accidents, 19.8

V

VEHICLE TRAFFIC, ILLUSTRATIVE ASSESSMENT, 5.4
VENTILATION PLANT, LOCAL EXHAUST, 9.28
VIBRATION
control of, 1.21
effect on new or expectant mothers, 3.13
noise generation, 10.18
VIOLENCE AT WORK, 19.7
definition of, 19.7
emergency procedures, 19.7
safe systems of work, 19.7
statistics on, 19.7
types of, 19.7
workplace, away from the, 19.7

W

WEBSITES
HSE (Health and Safety Executive), 4.4
WORK AT HEIGHT *see* **HEIGHTS, WORK AT**
WORK EQUIPMENT
children, 3.6
WORKING CONDITIONS
mothers, new or expectant, 3.10
WORKPLACE EXPOSURE LIMITS, 9.15
WORKPLACE SAFETY, STATUTORY PROVISIONS, 5.4
WORKPLACES, DANGEROUS, 3.6
WORKSTATIONS, DSE
assessment, 12.4
assessment records, 12.14
daily work routine of users, 12.6
definitions, 12.2
exclusions, 12.3

WORKSTATIONS, DSE – *Contd*
eyes/eyesight, ... 12.7
information provision, ... 12.9
new or expectant mothers,
effect on, ... 3.13
requirements for, ... 12.5
risk reduction, ... 12.4
shared, ... 12.17
training provision, ... 12.8

Y

YOUNG PERSONS *see* **CHILDREN AND YOUNG PERSONS**

Z

ZONES, EAR PROTECTION, . . . 10.11, 10.26, 10.28, 10.29, 10.30